W9-AEZ-756

FRESH. EGGS
Daily

FRESH EGGS Daily

raising happy, healthy chickens... **naturally**

LISA STEELE

st. lynn's
press

PITTSBURGH

Fresh Eggs Daily
Raising Happy, Healthy Chickens… Naturally

Copyright © 2013 by Lisa Steele

All rights reserved. No part of this book may be reproduced, stored, or transmitted in any form without permission in writing from the publisher, except by a reviewer who may quote brief passages for review purposes.

ISBN-13: 978-0-9855622-5-0

Library of Congress Control Number: 2013941486
CIP information available upon request

First Edition, 2013

St. Lynn's Press . POB 18680 . Pittsburgh, PA 15236
412.466.0790 . www.stlynnspress.com

Book design – Holly Rosborough
Editor – Catherine Dees
Editorial Intern – Marguerite Nocchi

Photo Credits: All photos © Lisa Steele

The following trademarked products are mentioned in this book: Bach Rescue Remedy for Pets, Blu-Kote, Brinsea EcoGlow, Dawn, Gatorade, Knox Gelatin, Kocci Free, Nutri-Drench, Pedialyte, Sevin, Verm-X, Vetericyn, VetRx for Poultry, Vitamins & Electrolytes

Disclaimer
Any medical or preventive chicken care advice offered herein is purely informational and not meant as a substitute for proper veterinary care. In cases of serious illness, the services of a qualified veterinarian should always be sought.

This book was printed in the United States of America on triple-certified FSC, SFI and PEFC paper using soy-based inks.

This title and all of St. Lynn's Press books may be purchased for educational, business, or sales promotional use. For information please write:
Special Markets Department . St. Lynn's Press . POB 18680 . Pittsburgh, PA 15236

10 9 8 7 6 5

$\mathcal{T}$HIS BOOK IS DEDICATED
TO MY GRANDMOTHER,
WHO PASSED DOWN TO ME
HER LOVE OF CHICKENS
AND ALL THINGS COUNTRY.

TABLE OF CONTENTS

INTRODUCTION

I grew up across the street from my grandparents' chicken farm in New England and kept chickens as a child. But the decision to enter into the world of backyard chicken keeping as an adult happened quite by accident. After college, I worked on Wall Street for a number of years, then got married. Shortly after that, my husband retired from the Navy and we moved to Virginia, where we bought a small farm. Because my husband loves horses, he wanted to keep some on our property. I, meanwhile, thought it would be fun to raise goats, since it appeared my days as a career woman were over and I needed something to channel my energy. I kept pestering him about getting a few goats. Finally, in an effort to pacify me, he suggested chickens – I guess thinking they would be less trouble.

Now, I have to admit that the idea wasn't all that appealing to me initially. My family always kept a few chickens in a shed out back while I was growing up. I remember wearing oven mitts to collect the eggs so I wouldn't get pecked, always worrying about being attacked by one particularly mean rooster named Bojangles. And the hens weren't always very nice, either. In hindsight, the chickens were probably just broody and hadn't been hand raised or paid much attention, but regardless, my experience with chickens to date hadn't been all that positive. However, not being one to turn down the chance to raise any type of animal, I quickly agreed to my husband's "compromise," figuring I would still work on him about the goats.

Long story short, we went to the feed store, picked out six chicks, talked a bit with the owner about what we would need for them over the next few days, and brought home our new family members. And that was the start of our journey. I immediately began a crash course, reading as much as I could about raising backyard flocks, barely keeping one step ahead of them.

Today's backyard chicken keepers name their chickens and consider them pets, not livestock.

But I guess chicken keeping is in my blood, because not only was it love at first sight, I had clearly been watching, listening and absorbing a lot more than I realized growing up. I just seemed to know instinctively what to do. I took to the chickens like a duck to water and never looked back. We raise them for the eggs only. I could no sooner eat one of our chickens than eat our cat or dog. Even after the hens stop laying, they will continue to be loved, fed and cared for until they have lived out their natural lives. In fact, older hens make wonderful mothers, teaching baby chicks the ropes.

While I'm sure my grandparents enjoyed their chickens and no doubt treated them well, their chickens were purely a food and income source, for the eggs as well as the meat. My grandmother lived to be almost 100 years old, and in the years prior to her death she delighted in sitting talking with me about her (and my) chickens. She couldn't quite understand my naming them all, taking photos of them, or whipping up homemade treats for them; and she just shook her head when I told her I had put up curtains in the nesting boxes, but all I have to say to this new generation of chicken keepers is: "These aren't your grandmother's chickens!" I know she would be amazingly proud of all I have accomplished and how many people I have helped with advice and suggestions on raising backyard chickens.

Though our methods might differ a bit on the finer points of chicken keeping, when I first started raising chickens, I knew I wanted to raise them like my grandparents did, as naturally as possible. Not only would it be healthier for them, it would be healthier for our family, since we would be eating the eggs. A big part of raising our own chickens is being sure we know what goes into them and how best to care for them – so using antibiotics, wormers and other chemical-laden commercial medications was never on my agenda.

Live long and prosper

Not a lot has been written about the benefits of herbal and other natural preventives relating to chickens. Few studies have been done and little has been researched. Prior to this new trend in backyard chicken keeping, most laying hens didn't make it much past their third birthday. When their prime laying days were done, they went into the stew pot and a new chick was bought for a few cents, raised, and the cycle continued. If one got sick, it was culled and replaced with a new chick. Life went on.

Today, backyard chicken keepers are treating their chickens as "pets with benefits," enjoying the fruitful laying years of their hens but continuing to raise, feed and love them well past their productive years. A chicken can live ten or twelve years or more, barring an unfortunate end at the hands of a predator, so keeping our chickens healthy and happy for the better part of a decade became my goal.

I started sifting through the Internet and reading every book about raising chickens that I could get my hands on. I found loads of conflicting advice on so many aspects of chicken keeping, but soon realized that the old-timers' advice and old wives' tales were the things that really spoke to me.

One thing I realized early on: Herbs are beneficial in most every aspect of the life of a flock, and central to maintaining their health and happiness.

What's so good about herbs?

Culinary herbs have numerous health benefits as well as wonderful aromatic properties. If you garden, you most likely grow some already, so adding a few more varieties to your plot each spring and incorporating their use into your chicken keeping can be a fairly seamless transition. Using herbal remedies on your flock is not only simple and inexpensive, it comes with virtually no potentially dangerous side effects or withdrawal periods during which you can't eat the chickens' eggs – and can be extremely beneficial to your flock. (See my listing of the health benefits of common herbs and flowers at the back of the book.) If you are raising your flock organically, then herbal preventives and remedies are invaluable.

Herbs are easy to grow. They don't take up much space (although mint does like to wander) and aren't picky for the most part about soil type. They don't need much attention to thrive, and there are countless varieties to choose from. Some herbs might even be perennials, depending on your climate, and come back each year. Others can be grown on your windowsill all winter. I have been growing herbs for as long as I can remember. Even when I lived in a tiny little apartment after I graduated from college, I always had a row of small flower pots filled with a variety of herbs on my kitchen windowsill: basil, tarragon, parsley and rosemary, to name a few. Since moving out to the country, I have built a raised bed just for herbs. Each year I add a few new varieties in addition to my standards and the many perennial plants that reappear year after year.

So when we began keeping chickens, it was only natural for me to start researching how to incorporate herbal remedies and preventives into raising a happy, healthy, holistic flock. It turns out every culinary herb (as well as many common flowers, such as roses, nasturtiums, bee balm) has amazing health benefits of some kind, whether your hens eat them, brush against them, or even ignore them, letting the herb act as an insecticide in the coop. Our chickens love to eat fresh parsley, basil, cilantro and mint. Herbs contain an abundance of nutrients and the chickens seem to know. It has recently been found that chickens can actually choose to direct nutrients either toward themselves or to their eggs; the more nutrients they eat, the more excess nutrition they have to direct into their eggs. Their eggs will be more nutritious and be healthier for you, as well.

While some herbs can be dangerous if used incorrectly, the culinary herbs are safe, so those are the ones I concentrate on. Because they are all edible, I can use them fairly loosely in no set amounts without worry of overdose. Birds in the wild don't rely on measured amounts, but instead seem to know instinctively which herbs and weeds they need – how often and how much – so I allow our chickens to make the same choices. I figure they do know best, and I always feed herbal remedies "free-choice."

Building strong immune systems

Chickens are masters at hiding signs of sickness and injury. Centuries of being very low on the food chain have taught them that any appearance of weakness or illness can lead to being targeted by predators or

In addition to using herbs in cooking, culinary herbs take center stage in holistic chicken keeping.

pecked by their flock mates. As a result, by the time even the most observant chicken keeper notices something is wrong, it can be too late. What to do? As I mentioned, there haven't been a lot of studies done on chicken health and there aren't many commercial medications formulated specifically for poultry. Add to that, if a chicken does appear to be sick, it can be nearly impossible to find anyone to treat her – and we know that many diseases spread quickly through a flock. The primary purpose of this book, therefore, is to help others *avoid* illness in their flocks in the first place.

Maintaining a thriving herb garden is far less expensive than paying for a vet visit, even if you are fortunate enough to be able to find a local vet who treats chickens. Growing herbs and using them as preventives and remedies is also less expensive than buying commercial products – so with not a lot of cost involved and a sometimes unknown (or not scientifically proven), but potentially positive benefit, I don't see any downside to treating your hens to a holistic diet and environment in which to live. Culinary herbs added to both your flock's and your own diet will lead to better health, stronger immune systems and a more aromatic, flavorful life.

Seeing first hand how healthy our chickens are and how easy it is to incorporate the herbs into their upkeep, I now grow a far wider assortment of plants than ever before, both to use in cooking and in conjunction with raising our chickens.

🐦 *What About Essential Oils?* 🐦

Fresh herbs are best, but dried are okay if you don't have fresh. I generally stay away from essential oils because they are extremely concentrated and some can be dangerous in large doses. While an overdose with any harmful side effects is quite unlikely, it's possible that your hens could have a reaction to large quantities of an essential oil. So stick with fresh herbs whenever possible.

My healthy, happy, aromatic flock

In the five years we've been raising chickens, we have never lost a single one to disease, never had a case of intestinal worms, no lingering parasites, no scaly leg mites, no pecking issues or impacted crops – all of which plague so many backyard flocks. Most importantly, we've never had hens with any of the recurring respiratory problems that are so prevalent in chickens. And despite my not vaccinating newly hatched baby chicks or feeding them medicated feed, we have not had a single one contract coccidiosis or Marek's disease.

That's good enough evidence for me that this holistic stuff works. It worked for my grandmother and her mother before her, and it works for me.

This book is the result of years of my own informal research and in-the-coop "testing" to show you how you can keep your chickens healthy and productive without the use of chemicals, medications or harsh commercial products, and how to have some fun while you're doing it. My book is not meant to be a beginner's guide to keeping chickens, although some basic care is included throughout, but instead assumes that you have the basics down and now you're ready to ramp up your chicken keeping and invite herbs and other natural treatments into your routine.

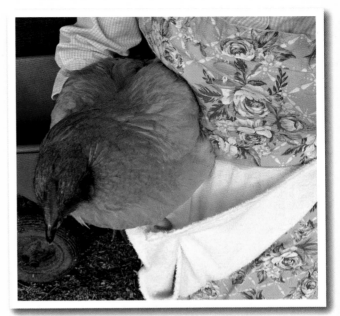

Spending quality time with each of your chickens enables you to spot any potential health issues early.

I wrote *Fresh Eggs Daily* to share with you what I've learned from caring for my own flock. In the chapters that follow, you'll discover how to incorporate both fresh and dried herbs into your chickens' feed, coop and run area. You'll learn why herbs are so important for baby chicks – and much more, to ensure a healthy, contented flock. You don't need to use all of my recommendations to make a difference. Take from this what works for you. Every little bit you do will be beneficial. And please let me know how you and your flock are doing.

Ready to get started? Grab a pair of kitchen shears and a basket and let's head out to the garden on our way to the coop!

ONE

IN THE COOP

nce you have made the decision to start raising a backyard flock of chickens, your first thoughts will most likely turn toward acquiring a coop for them. Coops come in all shapes, sizes and materials. You can build your own, buy a kit or buy one turn-key; you can convert a shed or playhouse. Personal preference comes into play here as well as the eventual number of hens you think you will want (do some serious thinking here, because one of the most oft-mentioned regrets I hear from fellow chicken keepers is that they didn't build their first coop large enough).

All right, so you have decided how many chickens you are going to want, and researched how many your municipality will allow you to have; now you need to multiply that number and then multiply it again because, take it from my personal experience, chickens are addictive. You can't stop at just three... or five... or even a dozen. You will see a pretty hen and want to know its breed and then decide you need one. You will want blue eggs and then green eggs, pink eggs and cream-colored eggs.

You will enjoy them so much and feel such a sense of calm and well-being watching them scratching and exploring your backyard, that you will want to keep expanding your existing flock. Each spring, chick fever will set in and you will be itching to bring home a few adorable, fluffy baby chicks from the feed store, order rare breeds online or even try your hand at hatching some. So plan ahead and design your coop for the flock you eventually want to have, not what you will be starting with.

There is no one perfect coop design. However, no matter what kind of coop you decide on, there are some basic design elements you will want to incorporate. Of course, safety from predators should be your main concern. Everything from raccoons to foxes, owls to your neighbor's dog will know immediately that you are raising chickens and decide they are on the menu. There are pros and cons to any coop style, but all good coops share certain characteristics. Your coop will need to have:

- Enough room for the number of hens you plan on raising
- Easy access for cleaning
- Adjustable ventilation
- Predator-proof latches on the doors and windows
- Roosting bars
- Nesting boxes

Size

The size of your coop will be your first decision. Rule of thumb calls for 2 to 4 square feet of floor space per bird, more for larger breeds, less for bantams and smaller breeds. You will need more space if you live in a colder climate and your hens will be spending a substantial time inside, although a smaller coop is easier for the hens to warm with their collective body heat – so you don't want to build a 10x12 coop for just three hens, for instance. But three have a way of turning into a dozen or more.

Ventilation

Good airflow is extremely important in your coop, especially in warm, humid southern climates – but even in northern climates in the winter. Chickens are highly susceptible to respiratory illness, and a coop without enough ventilation can cause a buildup of moisture.

This can lead to respiratory distress, as can the ammonia fumes created by decomposing manure, both of which can cause eye and mucous membrane irritation.

Your coop should be well-ventilated but not drafty, with the majority of the vents being above the level of the birds' heads when they are roosting. That creates airflow but not drafts. Vents allow dust and other particles to escape and allow fresh, oxygen-laden air in. Without adequate ventilation, carbon dioxide levels in your coop will rise and not enough oxygen will flow in to replace what the chickens breathe.

Always plan for more chickens than you originally start with.

Your coop needs plenty of windows and openings covered in ½-inch hardware mesh to keep predators out. One-fifth of your total coop wall area should be mesh-covered vents, with panels that can be opened and closed as needed, depending on the weather and season. Since respiratory illnesses are often the result of not getting enough fresh air, or of breathing moist, humid air, your coop needs to be ventilated year round.

Keeping out predators and other critters

Raccoons are adept at opening doors with many types of knobs, handles and deadbolts. Locking eye hooks, carabiners or even padlocks, are a much safer option. If your nesting boxes are situated with access from outside your coop, there should be a secure latch on the nesting box lid as well.

As soon as you build your coop, remember that you have also built a lovely new home for any resident rodents: warm, dry, filled with soft bedding, safe from predators and equipped with a nearby unending food and water source. Your coop is a field mouse

Chickens instinctively seek high ground for sleeping, to stay out of predators' clutches.

family's dream home! You want to discourage rodents from spending time in and around your coop. They can carry disease, injure or kill chicks or bantam breeds, and have been known to chew on the feathers and feet of unsuspecting sleeping hens, looking to supplement their diets with a bit more protein. They will even steal the feathers to make nests.

Any holes larger than an inch should be covered with ½-inch hardware mesh to prevent entry by mice. You should make regular inspections throughout the year, especially in the fall as the weather turns colder, to be sure no mice have taken up residence. But that's not the only thing you can do; see page 6 for a super herb that helps to discourage rodents.

Bedding

The floor of your coop should be covered with a thick layer of bedding to offer a soft landing from the roosts. There are three types of bedding I recommend: pine shavings, straw or shredded paper. There are also three types of bedding I can't recommend because they could actually harm your chickens. Let's start with what not to use.

Bedding materials to avoid

Cedar chips or shavings. These should never be used, as the cedar oils can be toxic and cedar itself is very aromatic and can cause breathing problems in chickens.

Hay. This should never be used because it is too "green," meaning it's not dry like straw and can quickly mold and harbor mildew or other bacteria.

Sand. While some people use sand on coop floors because it's easy to scoop out the manure, I don't recommend it. Not only is it not a good insulator in the cold months, it can harbor E. coli. Recent studies have shown that sand can contain and maintain a far greater E. coli infestation than even water can. Chickens may also be tempted to eat the sand, which would be detrimental for two reasons: They might be eating sand-covered poop inadvertently; and overeating sand can lead to impacted crops, as the sodden sand won't pass through their digestive system (see page 122 for more about impacted crops).

Safer bedding choices

Pine shavings. If you choose pine shavings, they should be the larger "chip" size, not sawdust size, which create too much dust. Pine needles can also be used; they have the added benefit of low-level antibacterial properties. Many people prefer them because, if you have access to pine trees on your property, the needles can be a free bedding choice. Just be sure to use only dry pine needles. Wet needles can get moldy, so better to collect nice, dry ones.

Pine needles can also be used and are economical.

Straw. One benefit of straw is that it is a superior insulator. Being hollow, it traps warm air both inside its shafts and in between them. Chickens also like to scratch through the straw to eat the chaff. In many areas of the country, straw is considerably cheaper than hay and makes for an economical bedding choice.

Shredded paper. If you work in an office and have access to a paper shredder trash can, shredded paper can be an economical, easy litter option. Most inks are soy-based nowadays so not harmful. Just be sure to change out your litter often, since wet paper can turn to paper mâché and you'll have chickens walking around trailing scraps of paper from their feet.

Fresh herbs for the coop floor

Whether you use straw, pine shavings or shredded paper on the floor of your coop, fresh herbs regularly strewn on the floor will make your coop smell fresh and also provide the hens with great health benefits. They will scratch around in the litter and rub against the herbs, allowing the essential oils to work their magic. The hens will also eat many of the herbs, which will provide them with added nutritional benefits. I try to choose the most aromatic and strongly scented herbs to use in my coop.

Mint is an obvious choice to add to your coop for both its rodent-repelling properties and the ease with which it grows. Mint thrives in moist soils and a wide range of climates.

Mint. This is the herb I use most in our chicken coop. Not only does mint grow like a weed all spring, summer and fall, it smells good and has rodent-repelling qualities. Liberally spreading fresh mint on the floor of your coop or planting a few mint plants outside around the coop can help keep rodents away.

Lavender. This is another great herb to sprinkle in your coop. Lavender is extremely calming and soothing, and hopefully helps instill in the chickens that the coop is a safe place in which to sleep, lay their eggs, take refuge when they aren't feeling well, or to seek out as protection from the elements or predators.

Oregano. Oregano is a very important herb that should be regularly added to your coop litter. Thought to combat coccidia, E. coli, Salmonella and infectious bronchitis, oregano is currently being studied by commercial poultry farms as an alternative to conventional medications.

Yarrow. Yarrow helps maintain respiratory health. Of course the best way to avoid respiratory issues is to keep your coop litter fresh and dry and change out wet litter quickly, provide plenty of vents and windows in your coop, and allow your hens access to the outdoors year round. But introducing some herbs that specifically target respiratory systems and sinuses is always beneficial.

Bundles of yarrow tied together and hung in your coop can help prevent respiratory issues. As an extra boost, add some rosemary and thyme, both of which also help with breathing. Especially in the wintertime, tying bouquets of these three herbs together at the stem end and hanging them upside down in your coop – so your hens can eat them at their leisure – can be beneficial in keeping mucous membranes, lungs and breathing apparatus in good working order.

A number of other herbs are beneficial to regularly introduce to your coop, including **bay leaves** (a natural insecticide), **lemon balm** (a member of the mint family and therefore a rodent deterrent), **lemongrass** (a fly repellent), and **pineapple sage** (extremely aromatic, and a good choice for the floor of the coop).

Refreshing the coop

I have developed an all-natural coop refresh spray that combines the rodent-repelling and insecticide properties of mint with soothing lavender in a white vinegar base (for both disinfecting and antibacterial qualities). Easy and inexpensive to make, a bottle stored in your coop and used any time your coop needs a bit of a refresh will help keep your coop sanitized and rodent free.

You can also substitute vodka for the white vinegar in the following recipes. Why vodka? Vodka is another fine natural cleanser. It not only repels insects, it actually kills them. It is antibacterial and kills mold and mildew. As an added benefit, vodka is odorless, unlike white vinegar, which has a distinctive smell.

Lavender Mint Coop Refresh Spray

What you'll need:
Handful of fresh mint
Handful of fresh lavender leaves and/or buds
Bottle of white vinegar (or inexpensive vodka)
2 one-pint canning jars
Spray bottle

What to do:
Divide the herbs between two canning jars, crushing them a bit with your fingers as you add each sprig to release the essential oils. Add white vinegar or vodka to each jar to completely cover, leaving at least ¼-inch headroom in each jar. Screw on the lids and then set the jars in your pantry or on the kitchen counter to age. Shake the jars every few days to mix the contents.

The mixture will start to turn a greenish-brown color and smell fragrant in a week or two, indicating that the spray is ready to use. Strain the contents into a spray bottle. Spray in your coop as needed to keep it smelling fresh.

Cleaning the coop

While a quick refresh results in a fragrant coop, take care that you are not merely masking any true odors that should be attended to. Any whiff of ammonia should be cause for a thorough cleaning. Chicken droppings emit ammonia and the fumes can irritate your hens' eyes and mucous membranes. Many chicken keepers will tend to reach for the bleach as a coop cleaner, but mixing ammonia and bleach, as you may know, can result in toxic fumes. A far better alternative is this natural orange peel and white vinegar coop cleaner.

White vinegar, as mentioned above, is a natural disinfectant with antibacterial properties. It kills mold and is an ant repellent. *Caution:* Although apple cider vinegar has wonderful health benefits, which will be discussed later in the book, it should *never* be used for cleaning, because it will attract fruit flies.

My coop cleaner includes orange peels. Citrus oil is a natural insect repellent and proven solvent, making it perfect for scrubbing your roosting bars. I add cinnamon sticks to the cleaner because cinnamon oil kills mosquito larvae. Mosquitoes can spread fowl pox, a disease that causes black spots on hens' combs. I also add vanilla beans to repel flies, mosquitoes and other insects.

As in the preceding recipe, vodka can be substituted for the vinegar in this recipe if you wish.

Orange Peel White Vinegar Coop Cleaner

What you'll need:

4 oranges (or 6 limes, 5 lemons, 2 grapefruit – or a mix)
2 cinnamon sticks
2 vanilla beans
Bottle of white vinegar (or inexpensive vodka)
2 one-pint canning jars
Spray bottle

What to do:

Peel the citrus and divide the peels between two canning jars. Break the cinnamon sticks in half and add to the jars. Slit each vanilla bean and add to the jars. Pour enough white vinegar or vodka into each jar to completely cover the peels, leaving at least ¼-inch headroom in each jar. Screw on the lids and then set the jars in your pantry or on the kitchen counter to age. Shake the jars every few days to mix the contents. The mixture will start to turn an orangish-tan color and smell fragrant in a week or two, indicating that the spray is ready to use. Strain the contents into a spray bottle to use full strength for scrubbing roosts or nesting boxes, or mix with water to scrub the coop floor and walls. The cinnamon stick and vanilla bean can be reused for a second batch, but use fresh citrus peels. (This cleaner also makes a wonderful all-natural non-chemical kitchen cleaner.)

HERBS FOR THE COOP

Bay leaves	Mint	Rosemary
Lavender	Oregano	Thyme
Lemon balm	Pineapple sage	Yarrow
Lemongrass		

TWO

IN THE NESTING BOXES

Nesting boxes are where your hens lay their eggs, or at least where they *should* lay their eggs. By providing boxes of the right size, filled with soft nesting material, in a quiet part of your coop, you will encourage your hens to use the boxes. If you are new to chickens, you can put plastic Easter eggs, golf balls, wooden or ceramic eggs – or even egg-shaped stones – in the boxes, to help entice your pullets to use the boxes when they start laying. If you already have chickens, the older laying hens will teach the younger new layers where to lay their eggs.

Making the boxes

Boxes should be roughly 12 inches square, preferably with a sloped roof so the hens can't roost on top of them and soil them, and with a low board across the front to keep the eggs and bedding from being kicked out. If the boxes are too large the hens will try to crowd into the same box to lay, so you want the boxes just large enough for one hen to sit and turn comfortably.

Nesting boxes can be purchased commercially or can be homemade. They can be wooden crates, wire baskets, wicker baskets, on the floor or raised off the floor, with a way for the hens to get to them.

Each nesting box should be just large enough for one hen to fit comfortably.

Rule of thumb dictates one nesting box for every three to five hens, but realistically, no matter how many boxes or how few hens you have, they are all going to want to use the same box and some rather animated and heated "discussions" will ensue. Why do they do this? One theory is that when a hen sees an egg in one of the boxes, she assumes that is a safe place for her to lay her own egg. Another thought is that the hens all have an innate desire to contribute to a community "clutch" of eggs, which one hen will then sit on once there are enough eggs collected to hatch a brood of chicks.

Nesting materials

A piece of rubber shelf liner cut to fit in the bottom of each box can help prevent broken eggs if the bedding does get kicked out or pushed aside. Appropriate nesting material includes straw, pine shavings, pine needles, shredded paper, shredded black-and-white newsprint or even soft fabric. Just before the hen lays her egg she will stand up, so a few inches of soft nesting material is critical to preventing the eggs from breaking as they are laid.

Privacy, please! Nesting box curtains

Hanging curtains over the front of your nesting boxes is useful as a way to encourage laying. Utilitarian curtains made out of burlap bags, feed bags, an old sheet or pillowcase, or fabric odds and ends can easily be stapled in front of the nesting box openings and will certainly work just fine. Or you can get more fancy and sew some custom curtains for your coop to hang on a curtain rod. Far more than just a frivolous flight of fancy, nesting box curtains actually serve several important functions.

The more protected and secure a hen feels while she is laying (and at her most vulnerable to predators) and the safer she feels her egg will be once laid, the more apt she is to lay eggs in the nesting boxes, instead of trying to hide her eggs elsewhere in the coop or yard. The curtains afford her privacy

Curtains hung on a rod are easily removed to be washed as necessary.

and a darker, more secluded place in which to lay. Curtains can also encourage a hen to go "broody" (i.e., decide she wants to sit on a nest of eggs for three weeks until they hatch), providing her with what appears to be a dark, secret place where she will be safe from predators' prying eyes. The curtains help her feel her chicks will be safe from the other flock members, as well.

Nesting box curtains can also help prevent egg eating, a hard-to-break habit. Once a hen starts breaking and eating eggs (either hers or eggs laid by other hens) and learns there's something tasty inside, it can be very difficult to get her to stop, and her bad habit can be picked up by other flock members. But by shielding your hens' view of freshly laid eggs, "out of sight, out of mind" can deter egg eating.

Laying an egg can be stressful for a hen. She needs to feel comfortable and safe while she is laying, secure in the knowledge that her precious egg will not be harmed and that she's chosen a safe place to lay her egg where it will not be disturbed.

Herbal aromatherapy for the nesting boxes

I believe that using herbal aromatherapy to create a soothing environment helps laying hens feel secure and relaxed. The first time I added herbs to our nesting boxes, one of our hens literally fell asleep in the box after she laid her egg! That was one relaxed chicken.

Of course you don't actually want your hens sleeping in the nesting boxes. That will lead to their defecating in the nesting material and result in dirty eggs, as well as squabbles with others who want to use that box. You want them to hop in, lay their egg and hop out to leave the box available for others to use. But it does make me feel good to know that our chickens feel safe and comfortable while they are laying. And of course if you have a broody hen and are hatching eggs, you are going to want to encourage your hens to stay in the boxes for the entire three week incubation period until the chicks hatch.

Lavender. Lavender is my first choice of herbs to use in my nesting boxes. It is extremely aromatic and will help to calm and soothe laying hens. Lavender is also an insecticide –

another beneficial property for nesting box use since the dark, moist area under the setting hens is a potential parasite breeding ground.

Mint. Mint helps to repel rodents and insects, so it makes sense to add some mint leaves to the nesting boxes as well. If rats are allowed access to your coop, they will roll the eggs out of the nesting boxes and steal them. Mice will also find nesting boxes appealing to build nests of their own unless you discourage them.

No matter what you use for nesting boxes, you can create a more calming environment with just a few handfuls of herbs and flowers.

Lemongrass (or citronella). This herb will repel flies that can bother laying hens during the warm months. It is a wonderful addition to the nesting box herbal blend. Flies can carry fowl pox, another reason that keeping your coop fly-free is so important.

Rose petals. Roses are not just pretty, they are also an antibacterial and they help to cleanse the blood of toxins – making roses a nice addition to the nesting boxes. And chickens *love* to eat rose petals.

Toss the herbs into your nesting boxes by the handful, or make up small sachets to nestle into the corner of each box.

Marjoram, fennel, nettles, parsley, nasturtium. All of these are purported to be laying stimulants, and therefore are obvious choices to add to your nesting boxes. Nasturtium has the extra benefit of being a natural wormer, and the chickens love the peppery taste of the nasturtium flowers, leaves and seeds.

Our coop has never smelled as good as it does now. Each morning I fluff up the straw in the nesting boxes, add a few more herbs as needed and enjoy a whiff of the fragrant blend.

Try tossing a few handfuls of mixed cut herbs into your nesting boxes and refresh them as needed. Your chickens will benefit from them, they look pretty, and you will enjoy how nice your coop will smell. My favorite combination includes **lemon balm, pineapple sage, rose petals, lavender, chocolate mint** and **thyme**, but feel free to experiment with what you have growing in your garden.

The chickens eat some of them, push some aside and happily nest in their boxes, for the most part leaving the herbs alone. I refresh the herbs every couple of days as they disappear or start to dry.

Coop to Kitchen
5 TIPS TO ENSURE CLEAN EGGS

1. Don't allow hens to sleep in nesting boxes
2. Refresh nesting material daily
3. Locate boxes across from your pop door
4. Discourage broodies to prevent squabbles over boxes
5. Collect eggs frequently

While the modern domestic chicken has been bred to lay an egg roughly every 26 hours, many factors go into that egg being laid. Stress, changes in routine, improper nutrition and illness can all affect how many eggs each will lay, so anything you can do to provide an optimal laying environment for your hens will result in greater productivity. Herbal aromatherapy and nesting box curtains are two easy ways to help create that environment.

HERBS FOR THE NESTING BOXES

Comfrey	Marjoram	Rose petals
Fennel	Mint	Seaweed
Lavender	Nettles	Thyme
Lemongrass	Parsley	

THREE

IN THE RUN

Your chicken run may be as large as your entire backyard for a large flock or as small as a pen attached to your coop for just a trio of chickens. Either way, it will most likely start out as a prettily landscaped, grassy area soon to be turned into a dirt-filled wasteland courtesy of many pairs of scratching chicken feet. Chickens not only eat grass and the roots of plants, they also scratch for bugs and worms and eat small pebbles to help them digest their food. You may be thinking it would be nice to let your chickens roam free all day, but unless you want your entire yard, along with your flower and vegetable gardens, devoid of anything living, it's best to confine them to a specific area.

Enclosing the run

An enclosed run is safest, for many reasons. Predators such as foxes, raccoons, snakes and neighbors' dogs are all intent on doing your chickens harm, and wild birds and rodents can carry disease and try to eat your chicken feed. So every attempt should be made to keep them out of the run.

Fencing should be welded wire, preferably with holes smaller than one inch. Chain link wrapped in chicken wire is also an appropriate fencing choice. It should be sunk into the ground several inches to deter digging predators. The run should be covered, either with poultry netting or chicken wire. This will prevent hawks and other raptors from gaining access from overhead and keep climbing predators such as foxes, raccoons and feral cats from climbing over – and keep your chickens from flying out.

Landscaping the run

Although it may seem like an exercise in futility, it actually IS possible to landscape your chicken run. A nicely landscaped run serves many purposes. The first is that the more aesthetically pleasing the run area, the more time you will want to spend there with your chickens – and that is beneficial to you both. Bushes and shrubs provide shade and wind protection, as well as help shield your flock from the prying eyes of both predators and neighbors (who might be less inclined to complain if the area is neatly appointed). Your flock will also be afforded a greater sense of security with cover should an errant hawk swoop by for a look. Another plus: a row of shrubs will act as a noise barrier to muffle the crowing of a zealous rooster and the jubilant "egg song" of your clucking hens.

Landscaping your run does take some planning and research.

An additional benefit to landscaping is the bugs that the various shrubs and bushes attract for the chickens to eat. Insects are nutritious and the chickens will keep busy catching and eating them.

There are several ways you can keep your run not only looking nice, but also turn it into an oasis for your chickens, while at the same time repelling both pests and predators.

The list of toxic plants and bushes is extensive, so research should be done before you decide what to plant in or near your run. The challenge, of course, is finding things to plant that the chickens won't eat and that aren't toxic to them. That can prove difficult but not impossible. For the most part, chickens do know what is good to eat and what's not – and they will stay away from anything that is dangerous unless they have nothing else to eat. But I always err on the side of caution and would never knowingly plant anything within their reach that could be harmful to them.

Some plants are poisonous for chickens and should be avoided at all costs. The list is exhaustive; please refer to page 137 for a list of common toxic plants and flowers – and do some research on your own before choosing what to plant if not specifically mentioned as safe here.

So what *can* you plant in your run? I have a few suggestions that have worked very well for me.

Rose bushes. Roses have turned out to be wonderful in the run. The chickens still strip the bottom two feet or so of leaves, but if you can cage the bushes until they are taller, the bushes will flourish and the chickens will love eating the fallen petals – and the rose hips. Once the petals have fallen off the flowers, I crack open the rose hips with my fingernail and give the chickens a vitamin-rich treat.

Butterfly bushes. This is another one of my favorites, perfect in the run. They are so pretty, come in many colors and grow fast, providing nice shade in the summer under their drooping branches. The chickens don't eat them and the flowers attract lots of bugs in addition to the butterflies.

Juniper. Juniper doesn't provide quite as much shade as the taller, sprawling, butterfly bush, but it adds year round color to the run, blocks predators' view – and the chickens don't touch it.

Hawthorn. This is a great choice. The berries are edible and drop in the fall, and so far our flock hasn't bothered the bushes or eaten any of the leaves.

Protecting your plants

Chickens will eat nearly anything you decide to plant, so small bushes and shrubs need to be caged until they are at least two feet tall and are well established. Additionally, placing stones, bricks or pavers around the roots of shrubs and trees can help protect fragile roots from being scratched up and disturbed.

Caging young plants helps ensure they can have a chance to grow.

Each spring I section off a sunny part of my run and plant grass seed, vegetables and some herbs. I keep the chickens out until everything has had a chance to establish itself and is mature and then they have quite the feast (just be sure that the seeds you plant are not coated or treated with any chemical fertilizers). The chickens end up scratching it all right back down to the dirt, but it looks pretty while it lasts and keeps the chickens occupied while providing them lots of nutrition.

A salad bar for chickens

Building a "salad bar" inside the run, with a wooden frame and a wire mesh top, is a great way to provide your chickens with grasses or other seeds to eat. The wooden frame keeps the chickens from scratching up the roots and only allows them to eat the treats once the plants mature and start to poke through the top screening. Wheatgrass is a particular favorite.

Plants for outside the run

Squash, cucumber, pea and **grape vines**, as well as **melons** and **roses**, can be planted alongside the run and trained to grow up the side of your run fencing. These will provide shade during the warm months of the growing season. Your chickens will also love pecking at the leaves and blooms that poke through the fencing.

Certain herbs are very beneficial for your run. Planting **mint** and **lavender** around the perimeter can help deter rodents, spiders and other insects. If you plant it *inside* the run, the chickens will eat it, which is good for them but defeats the purpose of it as a rodent deterrent. **Cayenne pepper** sprinkled around your coop will also help deter rodents, as will planting **catnip**.

Strategically planting various hebs around your run will help keep the insect population down.

Homemade remedies for natural pest control

Warm weather heralds the arrival of bugs and other pests, drawn to your chicken yard by the allure of free food, free-standing water and chicken manure. While chickens will eat many of the bugs that they come in contact with, other types they won't. Bad bugs and parasites can not only carry disease to your flock through contamination of the feed but some also transmit harmful diseases through direct contact. Others can damage wood in coops and barns. However, using pesticides could obviously be harmful to your chickens, so before you start spraying chemicals, consider these homemade natural remedies to repel pests instead.

Safe, easy and inexpensive, there's no downside to giving these ideas a try. And you already probably have everything you need to battle summer pests right in your garden, kitchen cabinets and pantry.

To repel flies: You can help reduce the fly population by planting **lemongrass, basil, dill, rosemary** or **mint** around the run area, as well as sprinkling food-grade diatomaceous earth (see page 31) around the feeders. Here's another favorite repellent: Soak several cotton balls in vanilla extract in a small canning jar, add fresh basil and mint leaves and cover the top with cheesecloth. Place in your coop or run – wherever flies are a problem.

Homemade fly catchers are easy to make out of empty wine bottles. Hang them in your run near your feeders. To make them yourself, and for my awesome new fly spray recipe, check out the next two pages.

To repel ants, earwigs, silverfish, fleas, cockroaches, millipedes and centipedes: Sprinkle food-grade **diatomaceous earth, black pepper, cinnamon** or **cloves** in your run. For around the run, planting **lavender, mint** and **basil** also helps keep them at bay, as does using a homemade spray of **white vinegar, cinnamon stick** and **garlic clove**. (Place a whole, peeled fresh garlic clove and a cinnamon stick into a spray bottle, fill with white vinegar and use to spray areas where ants seem to congregate.)

To repel spiders: Spiders are not fond of **spearmint** or **peppermint**. Nor do they enjoy **garlic cloves, citrus peels, lavender** or **white vinegar**.

To repel ticks: Plant **lavender** or **rose geranium** around the run or add some guineas to your flock. Guineas love to eat ticks.

To repel mosquitoes: Mosquitoes don't like **marigolds, fennel, rosemary, lavender, bee balm, basil, catnip** or **thyme,** so planting these around your run can reduce the population.

To repel snakes: Snakes don't like **marigolds,** either. Sprinkling **agricultural lime** or **sulphur** around the perimeter of your run or spraying **clove** or **cinnamon oil** can also help keep snakes at bay.

For flying insects, we have a bug zapper hanging over our run and each morning in the summer when I let our chickens out, they run right over and stand under the zapper waiting for me to clean it out to see what tasty toasted treats will fall to the ground!

All-Natural Fly Spray

Flies are a given around your chicken yard, attracted by the feed and manure as well as any standing water you may have. They carry disease, so you definitely want to discourage them. This all-natural fly spray is perfectly safe to use around your chickens to repel flies. The spray incorporates the antibacterial and disinfecting properties of white vinegar, with a citrus scent that is particularly unappealing to flies. Mint and basil are two other scents that flies try to avoid, so they are important components of the fly spray.

What you'll need:

1 lime
Fresh mint leaves
Fresh basil leaves
White vinegar
Pint canning jar
Spray bottle

What to do:

Cut the lime in half and place in the canning jar. Add a handful each of fresh-cut mint and basil leaves. Fill to ¼ inch from the top with vinegar. Then set the jar in your pantry, in the cupboard or on the kitchen counter to "age" for a week or two. Shake the jar every few days to reinvigorate the contents.

When ready to use, strain the contents into a spray bottle. Spray liberally wherever you see flies congregating in your coop or run.

Wine Bottle Fly Catcher

What you'll need:

An empty wine bottle
Painter's tape
Black spray paint
Short piece of thin wire
Fishing line
Raw meat or fish scraps
Maple syrup
White vinegar

What to do:

Tape around the body of the wine bottle, just under where the curve ends. Spray paint the top of the bottle black, above the tape (this confuses the flies, rendering them unable to fly back out once they fly in). When dry, peel off and discard the tape, then twist the wire around the neck of the bottle and make a small loop so you can hang the bottle. Attach as much fishing line to the loop as needed for where you plan to hang the fly catcher.

Drop some pieces of raw, cut-up meat or fish into the bottle, along with about ½ cup of maple syrup and a generous splash of vinegar (the vinegar will deter bees but still attract flies, wasps, yellow jackets and hornets).

Hang the fly catcher in your run, preferably over the feeding area. The premise is that the flies will fly in, attracted to the stinky, decaying fish and the sugars in the maple syrup, then get stuck in the syrup and not be able to fly out again.

It is possible to have a beautifully landscaped run that is functional as well as eye-catching. A few low stone walls make certain areas visually pleasing, protect your plant roots and also provide a place for your chickens to get up out of the mud or snow. A variety of bushes and shrubs will make your run look nice while providing shade, a supply of bugs and insects and a predator screen. Adding a few herbs around the run will help to keep the bad bugs and pests out. The joy that a pretty chicken yard brings me makes it worth the extra effort.

Placing bunches of hanging herbs is attractive and also makes a great pest deterrent.

HERBS FOR IN AND AROUND THE RUN

Basil	Fennel	Mint
Bee balm	Garlic	Rosemary
Catnip	Lavender	Thyme
Cinnamon	Lemongrass	
Citrus peels	Marigolds	

FOUR

IN THEIR DUST BATH AREA

Dust baths are a chicken's way of keeping clean. The fine sand or dust in their bathing area keeps their feathers clean and helps them stay free of mites, lice and other parasites. Dust baths are not just an enjoyable experience for your chickens, they are critical to keeping them parasite-free. More than that, chickens not allowed a place in which to bathe can actually start pecking at each other and suffer stress.

Your chickens will enjoy bathing and sunning themselves on nice days year round. You will enjoy watching their antics as they squirm and twist their bodies. Of course, the first time you catch them bathing you might think they are having convulsions or writhing in pain – and the first time you spy a bunch of sunning hens, wings spread, necks twisted and legs all a-kilter, you will have to look twice to be sure they aren't all victims of a mass predator attack.

A good "bath" entails the chickens' digging a shallow bowl in the earth with their feet to loosen the dirt and then rolling and wriggling around in it, working the dirt into their feathers and skin. They will happily toss dirt onto their backs, flap their wings and even roll their heads into the dirt to ensure every bit of their body is covered. Bathing seems to be a social event and you will often see a few bathing at the same time. When they are finished, they will loll contentedly in the sun, enjoying the warmth and absorbing Vitamin D.

When they are ready to get up, they will shake all the dirt out and then preen themselves to redistribute the oils into their feathers.

The best dust bath ever

Although your chickens will most likely choose their own preferred bathing area, you can help make bathing easy for them by finding a nice dry spot in the run, preferably covered so it doesn't get wet, where there is fine dirt or sand. Then add some **fireplace ash,** food-grade **diatomaceous earth** (see next page) and **dried herbs.** You can also try filling a large shallow tub or bowl, tire inner tube, children's sandbox or other receptacle with dry dirt or sand to encourage them to use it.

Kiddie pools make great receptacles for dust baths.

Wood ash/fireplace ash. This is a wonderful addition to the chickens' dust baths. Make sure to only use wood fireplace ash, not briquettes or any wood that has been treated in any way or had lighter fluid or other chemical coating on it. Charcoal wood ash contains Vitamin K (a blood clotting agent), calcium and magnesium. Charcoal absorbs toxins and research suggests that after forest fires, wild animals will consume it for its medicinal properties. Charcoal is also a natural laxative and removes impurities from the body, as well as any worms. Chickens can greatly benefit from the charred wood in their dust bath; they will nibble pieces of it as they bathe – much like charcoal pills for humans.

Food-grade diatomaceous earth. This amazing stuff kills mites, lice, fleas, ticks and other parasites. The DE filters through the hard-shelled bodies of the parasites with razor-sharp microscopic silicon particles, causing the parasites to become dehydrated, blocking their airways and attacking their respiratory systems. Add this to the dust bath area, and while the chickens bathe and wiggle in the dirt, the DE will make its way into and underneath their feathers, suffocating any parasites.

🐦 A Few Words about Diatomaceous Earth 🐦

It is always best to wear a mask while using DE. Though non-toxic and harmless to mammals when eaten, inhaling the microscopic particles in the dust can cause lung irritation in humans and chickens. Some experts advise not using DE in dust bath areas because of the possible risks, but I feel that any as-yet unproven possible risk is far outweighed by the very real chance of your chickens contracting mites and then possibly having to be treated with Sevin dust or another equally harmful carcinogen or commercial preparation – which I DO NOT recommend. Mixing the DE with the wood ash can help prevent excess inhalation of the DE by your hens. But as a precaution, be sure to wear a mask and take care when applying DE in the dust bath or coop area. And don't allow your chickens into the vicinity until the dust settles.

Dried herbs. Sprinkled in the dust bath area, dried herbs can be extremely beneficial and help prevent mites and other parasites. Dried **lavender, mint** and **rosemary** are natural insecticides. Dried **anise, dill, fennel, ginger, mint** and **seaweed** are excellent aids in disease and parasite prevention. Dried **wormwood** (artemisia) is thought to keep lice and mite away. Dried **yarrow** is an anti-inflammatory and helps to clear respiratory systems, as do **thyme** and **rosemary**.

Unless your chickens miraculously decide that they like where you have chosen for them to bathe, your run will soon look like a minefield with small depressions and craters all over.

Dust baths seem to be a community event.

Start 'em early

You will want to start your chickens off with good hygiene habits at a young age. I fill a small plastic tray with sandy dirt and put it in the brooder box for my new chicks. Even week-old chicks will hop into the tray and squirm around and flap their wings. It's healthy for them, keeps them busy and they are adorable to watch. As an added benefit, the dirt doubles as the "grit" they need to digest their food.

As the chicks get older, their bath gets bigger. When they are finally old enough to spend time outdoors, the bath goes outside with them.

I recently hatched a handful of chicks under a broody hen and began letting them out to explore a bit when they were about a week old. One of the very first "field trips" the mother hen took them on, she stopped to "bathe" as they watched. Within a few days, the chicks were bathing on their own.

Dust bathing seems to be one of my chickens' favorite pastimes. Encourage yours by creating an environment where they can bathe in comfort. They will thank you. However, mites, lice and fleas will not.

HERBS IN THE DUST BATH

Anise	Lavender	Wormwood (artemisia)
Dill	Mint	Yarrow
Fennel	Rosemary	
Ginger	Seaweed	

FIVE

IN THEIR FEED

What is the best type of feed for your laying hens? You have many choices. Within the category of layer feed there's organic, non-organic, non-GMO, soy-free, corn-free, whole grain, crumble and pellet, just to name a few – and of course many different brands. Some feeds are supplemented with probiotics, marigold, even grit, but most are not. Your choice really comes down to your personal preference, cost and availability in your area; any good quality layer feed will provide a fully balanced diet for your flock and will have adequate vitamins and nutrients, namely calcium, to ensure strong eggshells.

Mixing your own feed is also an option, but it can be fairly complicated to get the right balance. A discussion on mixing your own is well beyond the scope of this book and not recommended, especially for those just starting out with chickens.

Layer feed should be substituted for starter/grower feed at around 18 weeks of age, or when your chickens start laying. Regardless of the type feed you choose, there are a few natural supplements that can be added for your chickens' optimal health.

Customizing your feed mix

Breakfast is the most important meal of the day, for chickens as well as for people, and our chickens get a custom mix every morning before they get any treats. I developed what I call the "Breakfast of Champion Layers" several years ago and have been feeding it to our chickens as their main feed ever since. We have never had such beautifully-colored eggs with nice strong shells – or as healthy-looking hens – as we do now. All our chickens are gorgeous and glossy, with shiny feathers, bright eyes and rosy combs and wattles.

A full-grown layer hen eats approximately ½ cup of feed per day (more in winter, less in summer, and more if they are penned and don't have access to grass, bugs and weeds). I leave feed out all day for our flock, but alternatively you can feed in the morning and again in the evening, as much as your hens will eat in about half an hour. Either way, chickens self-regulate and won't overeat.

Breakfast of Champion Layers

1 50 lb. bag layer feed
1 large canister whole raw oats (grocery-store size canister)
4 cups shelled sunflower seeds
4 cups dried seaweed
2 cups flax seed
2 cups assorted dried herbs
2½ cups garlic powder (approximately 3% of the mix)
1½ cups probiotic powder (approximately 2%)
1½ cups food-grade diatomaceous earth (approximately 2%)
1 cup brewer's yeast

About the natural supplements in your mix

Oats. Oats are an excellent source of protein, fiber and antioxidants, and the chickens love them. Oatmeal contains the B vitamins thiamine, riboflavin, niacin and choline, plus calcium, copper, iron, magnesium, fluorine and zinc. According to the U.S. Department of

Agriculture's Farmers Bulletin, baby chickens that have a ration of oats added to their diet will be healthier than baby chickens that aren't offered oats.

Sunflower seeds. Sunflower seeds are packed with Vitamin B for cardiovascular health and Vitamin E, an anti-inflammatory. They are a wonderful source of antioxidants as well as magnesium and copper, both needed for strong bones.

Seaweed. Seaweed is rich in vitamins and minerals. When added to a chicken's diet it will result in a higher iodine level in their eggs and brightly colored yolks. Seaweed can increase fertility as well as overall health. Chickens will grow faster, especially when seaweed is fed in conjunction with brewer's yeast. Some research shows that feeding seaweed can result in reduced levels of cholesterol in your hens' eggs and increased levels of omega-3s. Seaweed is also a source of prebiotics, which, working hand in hand with probiotics, stimulate the growth of beneficial microorganisms in the digestive tract.

This tasty, nutritious feed mix will help keep your chickens in top shape and ensure the best quality eggs possible.

Flax seed. Adding flax seed to their feed will increase the omega-3 levels in eggs, making them more nutritious. The amino acids in flax seed also improve egg production, overall health and feather quality. And flax seed helps hens maintain a healthy weight. Studies have shown that adding flax seed to a diet reduces the chance of contracting ovarian, colon, breast and lung cancer in chickens (not just in humans!).

Garlic. You can add garlic to your chickens' diet in several different ways. You can float whole cloves in your waterer (mashed up a bit), replacing them every few days; you can offer crushed fresh cloves in a small dish free-choice (figure on one clove for every 10 to 12 hens); or you can add garlic powder to their feed. I have tried all three methods and find it easiest to just add the powder to their feed – but every once in a while I also give them a bowl of the fresh garlic or add the cloves to their water for an added punch of immunity.

Good-quality layer feed plus a few natural supplements will ensure optimal flock health.

The garlic helps repel fleas, ticks and other parasites, controls the odor of the manure and is a natural wormer. It has overall health benefits, resulting in a higher white blood cell count and supporting respiratory health and the immune system. Adding garlic to their diet will result in a better feed conversion ratio (the measure of an animal's efficiency in converting feed mass into increased body mass). While some say that the garlic will taint the taste of eggs, I have never noticed an off taste in our eggs, and I use a lot for baking, where it would certainly be noticeable.

Probiotic powder. Probiotics assure better intestinal health in your flock. They aid digestion, assist in nutrient absorption, help boost productivity and support the immune system in general. Hens who are fed probiotics maintain healthy weights and lay larger, better quality eggs with stronger shells.

Probiotics are also thought to help combat coccidiosis, Salmonella and E. coli in flocks. E. coli and coccidia have been shown to exist in virtually all poultry manure samples, but this only becomes a problem for a flock when their digestive environment is conducive to the bacteria reproducing to unmanageable levels – but not when intestines have plenty of "good" bacteria. Studies have shown that incidence of Salmonella in chickens can be reduced by 99% using a diet that includes probiotics.

Probiotics, such as those found in yogurt, are as beneficial to poultry health as they are to human health. However, excessive amounts of dairy products in their natural state can cause diarrhea in chickens. Their bodies aren't designed to digest the milk sugars found in dairy. For this reason, probiotic powder, which is specifically designed for poultry use, is a far superior choice to introduce good bacteria into their diet. The powder also has a much longer shelf life, as long as it is stored in a dry, airtight container.

"Probiotic" literally means "for life." It is the opposite of "antibiotic." Any time antibiotics are administered to your flock, probiotics should be given in conjunction, to promote the regrowth of good bacteria.

Food-grade diatomaceous earth. This is an all-natural, silica-based, crushed fossil that acts as a parasitic and kills hard-shelled insects. It is effective on both internal and external parasites, including fleas, ticks, flies, aphids, earwigs, silverfish, crickets, millipedes, centipedes and digestive worms.

DE sprinkled around the feeders controls flies and ants in the summer. Only food-grade DE should be used around the chickens, because they will inevitably end up eating some. And be sure that neither you nor your chickens are breathing the DE dust. Let it settle before letting chickens into the area, and keep a mask handy for yourself as you sprinkle, since the dust particles can irritate the throat and lungs.

DE added to chicken feed, according to an article in *Poultry Science*, increases shell weight and thickness, increases egg production and hens' body weight, and keeps bugs out of stored feed.

Brewer's yeast. Brewer's yeast is a very beneficial supplement, which contains all of the essential amino acids. And the niacin in the brewer's yeast helps grow nice, strong legs and bones. Brewer's yeast provides numerous other benefits in the form of better circulation and healthy nervous and cardiovascular systems.

Add a sprinkle of dried herbs

I dry various herbs and store them to use as feed supplements year round. After drying the herbs completely on wire cooling racks or my herb drying frames, I crush and mix them in a large bowl and then store them in airtight glass jars. I sprinkle a bit of the herbal mix into my chickens' daily feed.

Oregano. Studies are being done by commercial chicken producers that seem to point to oregano as a natural wormer and antibiotic. Adding some dried, crushed oregano to your feed might help guard against parasites, E. coli, Salmonella, coccidia and other bacteria in your chickens.

In addition, the following herbs can be mixed and matched into your own custom blend, depending on what you have available. Try to choose several herbs from each category for a well-rounded supplement:

Dried herbs can be mixed and stored for use through the winter.

Add dried herbs to your chicken's feed year round for optimal health benefits.

LAYING STIMULANTS
Fennel
Marjoram
Nasturtium
Parsley

RESPIRATORY HEALTH
Bee balm
Dill
Oregano
Thyme

GENERAL HEALTH
Cilantro
Sage/Pineapple sage
Spearmint
Tarragon

ORANGE EGG YOLKS
Alfalfa
Basil
Dandelion greens
Marigolds

Herb Drying Rack

Over the years, I have been enlarging our herb garden and growing more varieties of culinary herbs to use fresh all summer, and dry for use over the winter. This tiered drying rack is easy to make, can be made from old picture frames for an inexpensive project and collapses for easy storage when not in use.

The tiers allow you to separate three different herbs on the various levels, and screening allows for good airflow to ensure that the herbs dry quickly and evenly. The best part is that you can use old picture frames – and who doesn't have a stack of those in their garage? Even if you don't, you can pick up inexpensive framed artwork at a thrift shop or yard sale and repurpose the frames.

What you'll need:

3 wooden picture frames of varying sizes
Can of spray paint
Window screening (great use for
    discarded screen windows that
    have slight damage)
Length of chain (approximately 8 feet)
20 small eyehooks
Cordless drill with small drill bit
Pliers
Staple gun/staples
Scissors

What to do:

Spray paint the frames in the color of your choice and let dry.
Cut pieces of screen to fit the back of each frame, leaving ½-inch overlap, and staple around the edges, pulling the screen taut. Pre-drill a hole in the corner of each frame on the right side and also on the underside of the upper two tiers. Screw an eyehook into each corner.

Cut chain into 8-inch lengths, using the pliers to bend open each end. Attach one length of chain to each eyehook on the bottom tier, then attach the middle tier to the bottom tier, with the painted sides of each frame facing up. Attach a length of chain to each eyehook on the top of the top tier and then open the top link of each length of chain to connect the four. Attach one final shorter length to hang your rack.

Find a place out of the way (where it's not damp) to let your herbs dry, and hang your drying rack. The curtain rod in a spare bathroom or a laundry room rod works well. Different herbs take different lengths of time to dry depending on their water content, so keep checking them periodically. Some herbs, like **dill** and **parsley**, will only take a few days; others such as **basil** and **sage** might take a week or so – depending on the relative humidity in your home and how dry the air is.

When the herbs are completely dry, crush them finely and store them in covered glass jars, labeled with the name of the herb.

Providing grit and a calcium supplement

In addition to the layer feed, there are two essential supplements that your chickens need: **grit** (either commercial grit or frequent access to small stones, pebbles and coarse dirt); and a **calcium supplement** (either crushed oyster shells or crushed eggshells). These should both be fed separately, free-choice, so each hen can eat as much or as little as she needs.

Calcium from oyster shells and eggshells. You can purchase crushed oyster shells or simply save all your eggshells to feed back to your chickens. Farmers and homesteaders have been feeding eggshells to their chickens for hundreds of years. It makes sense. Why throw out something that is a free source of such an important nutrient that your chickens need? Laying chickens need a lot of calcium to ensure strong eggshells. If they do not have enough calcium to create the shell, they will start leaching the calcium from their bones for their eggs, which will result in very thin-shelled eggs as well as calcium-deficient chickens.

Feeding your chickens eggshells will not lead to egg eating. In fact, I think it does the opposite. By providing them a constant supply and access to as much as calcium as they need, their bodies won't crave any more and they won't be tempted to peck at the shells of the eggs they lay.

I have done "unofficial" side-by-side taste tests and find that my chickens will always eat the crushed eggshells before the oyster shells. They seem to prefer them.

Simply rinse your eggshells and remove the membrane, then let them air dry before crushing them into small pieces. *But don't pulverize them* – if the pieces are too small, not enough calcium ends up being absorbed into the chickens' systems.

The better diet your chickens have, the healthier they will be and the more nutritious their eggs will be.

FEED SUPPLEMENTS

Brewer's yeast
Diatomaceous earth (food-grade)
Flax seed
Garlic

Oats
Probiotics
Seaweed
Sunflower seeds

Eggshell or oyster shell
Grit

HERBS IN THEIR FEED

Basil
Bee balm
Cilantro
Dandelion
Dill
Fennel

Marigold
Marjoram
Nasturtium
Oregano
Parsley
Rosemary

Sage
Spearmint
Tarragon
Thyme

SIX

IN THEIR WATER

An egg contains approximately 75 calories, is a wonderful source of nutrients (including Vitamin A, choline, folate, calcium, potassium and phosphorus), and is made up of approximately 13% protein, 10% fat, 1% carbohydrates and 65-75% water. Given the fact that a chicken lays an egg roughly every 26 hours, you can understand why access to clean, fresh water is so important for chickens – maybe even more so than for any other type of animal. Laying hens drink more water than roosters and non-laying hens, nearly two times more. Water is drawn from the hen's body to form each egg, so she needs to replenish that water on a daily basis, as well as drink enough to sustain her own needs. Daily water requirements vary depending on the temperature and time of year, but on average, each hen will drink half a liter of water a day.

Water need not be provided overnight in your coop as long as feed is not there either. I don't leave food or water in our coop as a general rule, since chickens can't see in the dark well at all and don't eat or drink once they go to roost. As long as you are there to open up the coop at daybreak or shortly thereafter, providing food and water only in the run area is fine and will keep your coop drier, cleaner and less likely to attract bugs and rodents.

If a hen is denied a suitable water source for even a short period of time, she will stop laying eggs, so it is critical that clean, fresh water be made available to your flock at all times during their waking hours. Being deprived of water for just 24 hours can cause a reduction in egg production that can take up to three weeks for the hens to recover from; and being deprived of water for 36 hours can force them into a molt and cause them to cease laying for up to two months.

Contamination and attention to water temperature

If their water source runs dry and the chickens are forced to drink out of dirty or stagnant puddles, they can contract Salmonella, listeriosis, botulism or coccidiosis. If their water gets contaminated with feces or is allowed to get too warm in the summer, the hens will just stop drinking. That is extremely stressful to their bodies and will fairly quickly result in decreased egg production.

Fresh clean water is vital to your chickens' health.

I can't emphasize enough how important it is to provide ample water during the day, placed in the shade during the warm months and in the sun during the colder months. It is preferable to set out several waterers throughout your run to ensure that even those at the bottom of the pecking order are able to get a drink when they want; it's also backup if one waterer should run out, become soiled with feces or tip over.

Good things to add to their water

There are several things you can add to your flock's water to boost their health naturally:

Garlic. Garlic enhances immune systems and increases respiratory health. It is thought that mites, lice, ticks and other parasites are not as attracted to the blood of animals that eat a lot of garlic. Garlic is also a natural wormer and reduces the smell of chicken manure in flocks that are fed garlic regularly. Break up a clove into several pieces and add it to your chickens' water. The clove should be removed and replaced every few days with a fresh clove, but I find that the chickens end up eating them, which is fine also. I have been adding garlic daily to our chickens' water for several years and have never found that the garlic taints the taste of our eggs the least bit, as I mentioned earlier.

Garlic is a natural wormer.

Blackstrap molasses. You can supplement the water with small amounts for an added dose of vitamins and nutrients. While too much molasses can cause diarrhea in chickens, a small drizzle added to the water during times of extreme stress from heat or injury can work wonders.

Herbal tea. Another easy and beneficial way to introduce more herbs into your flock's diet is to brew them herbal tea. Either fresh or dried herbs can be brewed into tasty tea for your chickens to enjoy. Some of my chickens' favorites are **rosehip/ rose petal, basil** or **lemon balm** tea, but I also brew them **parsley, oregano** and many herbal blends of tea. **Dandelion** tea is also a healthy choice, since dandelion greens contain high levels of protein, vitamins and minerals.

Adding a variety of herbs or fresh fruit to their water makes a tasty treat.

To brew the tea, fill your teakettle or a large saucepan with water, add some fresh-cut or dried herbs, bring to a boil, then turn off the heat and let the tea steep for ten minutes or so. Let cool to room temperature. You can stir in a bit of honey for some great natural antibacterial health benefits as well. Once cooled, you can either strain the solids or leave them in for your chickens to eat.

Vegetable cooking water. For another nutritional boost, reserve your vegetable cooking water and let your chickens drink it. I never salt the cooking water when I boil fresh vegetables, making it perfect to offer to our flock.

Apple cider vinegar. The addition of apple cider vinegar is a way to make the water even more beneficial to chickens. Among its many benefits, it balances the water's pH, thereby

Muddy puddles can breed bacteria, so clean water is a must.

creating an environment that is inhospitable to microbes and bacteria. Studies have shown that the vinegar actually makes the water more palatable to hens and can also be used to encourage a sick or injured hen to drink more. Our chickens do seem to prefer the taste of the water with the vinegar in it. I have on occasion set out two water bowls, one plain water and one with apple cider vinegar, and they always seems to gravitate toward the bowl with the vinegar and drink more of it.

You should use only unpasteurized, raw apple cider vinegar that has the "mother" enzymes, such as Bragg. (The mother contains the health benefits that assist in overall flock health.)

The apple cider vinegar should be added to clean water in the ratio of one tablespoon per gallon of water, several times a week. (If you have a pH testing kit and can check the pH of the water, experiment a bit with different ratios. You are aiming for a pH of 4.0.) Do not add vinegar to metal waterers as it will cause them to rust. Plastic, rubber or stoneware waterers should be used instead any time you add vinegar to the water.

Adding apple cider vinegar to the water not only helps keep the waterers clean and free of bacteria by creating an unhealthy environment for algae and other "bad" bacteria, it also helps regulate your hens' digestive systems and increase the good bacteria in their intestines. It is an overall health and immune system booster and increases the absorption of calcium and other minerals. An antiseptic, apple cider vinegar kills germs that can cause respiratory issues. And it promotes healthy mucus flow. Importantly, it is thought to help combat coccidiosis/coccidia, which is present in nearly every chicken run in varying amounts and can be devastating to your flock if they contract it. This amazing vinegar can also help prevent yeast growth in the crop, a condition that can lead to sour or impacted crop, two potentially serious problems.

Adding apple cider vinegar to water is beneficial in many ways.

You can make your own apple cider vinegar easily and cheaply. The commercial brand can be expensive and if you add it regularly to the water, you will go through a fair amount; so by making it from apple peels and cores that would be otherwise thrown away or composted, you can save a substantial amount of money. There are several ways to make apple cider vinegar, but Mother Earth News provided what seemed the simplest method, which I now use. It's very easy, although it does take a bit of time, so you need to plan ahead.

Placing their water dish in a shady place helps keep water cool on a hot day.

Homemade Apple Cider Vinegar

What you'll need:

5–10 apples or more (preferably organic)

Large glass or stoneware bowl

Sugar (optional)

Canning jars (sterilized)

What to do:

Wash, peel and core the apples (you're going to be using only the peels and cores for the vinegar). There is no set amount of apples you need to make a batch. You can make as much or as little as you want, and you can even store the peels and cores in the freezer until you have enough.

When you have collected a nice stash of apple scraps, place the peels and cores in a large glass or stoneware bowl and cover with water by an inch or so. To help the fermentation/yeast process work faster, you can add ¼ cup of sugar for each quart of water you use and stir to mix thoroughly, but this step is optional. You can omit the sugar altogether; the process will just take a bit longer.

Place a heavy plate that is slightly smaller than the diameter of the bowl on top of your apples to keep them completely submerged in the water. Cover the bowl with a clean kitchen towel and let it sit for a week in a cool, dark location – a spare room, the garage or basement, for example. UV light will destroy the fermentation process, so dim light, at most, is recommended. Try to keep the bowl in an area that stays between 65 and 85 degrees, a good fermentation temperature range.

After several days, the mixture will begin to bubble and foam as yeast forms. After a week, check under the plate and spoon off any black mold that has grown. If you keep the bowl in a cool spot you shouldn't have any mold, but if you do, it's fine to scrape it off and discard it.

Now strain out the apple solids and pour the liquid into sterilized canning jars, leaving about an inch of headroom at the top of each jar. Discard the solids, or at least discard the cores. The peels can be fed to your chickens if you wish. But NOT the cores. Apple seeds contain traces of cyanide and should not be fed to chickens.

Cover each canning jar with a square of doubled cheesecloth and screw on just the ring part of the top. This allows the yeast to "breathe" and prevents the metal from corroding. Hang on to the flat parts of the lids – you'll need them later.

Store the jars on a shelf in your pantry, a cupboard or back in the garage or basement, and wait about six weeks. A film should start forming on the top of each jar. This is the "mother." You can open up the jars and stir or swirl them so the mother settles on the bottom, to encourage more to grow on top. After about a month, you should see definite mother growing on top. About the same time, the liquid will start to turn cloudy but still fairly light-colored, without a distinct "vinegar" smell. The color will begin to deepen during the next two weeks and you will start to see residue settling on the bottom of each jar.

At the end of six weeks, remove the cheesecloth, replace it with the flat part of the canning jar lid and screw the ring back on. There should be a distinct "vinegar" smell now and jellyfish-like masses floating in the jar.

By this point, the yeast will have eaten all the available sugars and you will be left with a "shelf-stable" vinegar. The flavor will develop and evolve over time. Stored in a cool, dark place, the apple cider vinegar will last indefinitely.

Note: If you save some of the mother from each batch and add it to the next batch, the vinegar will be finished more quickly. It's hard waiting the full six weeks for your first batch, but once you get going with making your own, if you start a new batch each week, you will always have a batch of homemade apple cider vinegar ready, going forward, for a fraction of the cost of the store-bought variety.

IN THE WATER

| Apple cider vinegar | Garlic | Molasses |

SEVEN

WHEN YOU WANT TO SPOIL THEM

We all want to spoil our pets, and none more so than the pets that lay us beautiful, fresh eggs. But too many treats aren't good for any of us, including our chickens – and in fact can lead to reduced egg production and health problems in your flock if you overdo it. The trick is to offer the right kind of healthy treats in the right amounts so you can feel good about "treating" your chickens the best you can. Most kitchen scraps can be fed to your chickens, but should be considered "junk food" – sorry. However, logically, the more healthy a diet your family eats, the more nutritious your leftovers will be for your chickens.

As a general rule, treats should make up only 10% of any chicken's overall diet. Treats should only be fed in the afternoon, once your chickens have had time to fill up on their balanced layer feed. After they have fulfilled their daily requirements, it can be treat time. These treats can include overripe, wilted or stale fruits, vegetables, grains, breads, meat scraps and other kitchen leftovers. Nothing should ever be offered that is moldy, rotted or too salty or sugar-laden, or is deep-fried.

"Green Treats"

The exception to the 10% rule is so-called "green treats," which they can have in unlimited amounts. Those would include grass, weeds, lettuce, dandelion greens, parsley, kale,

Swiss chard, berries and bugs, to name a few. Basically anything they would likely find in a free range environment is going to be fair game in unlimited amounts.

Weeds. Weeds are extremely nutritious, free and a big favorite of our flock. Over the years, I've picked many, many bucketfuls and offered them to our chickens to pick through. I have made it a point to watch closely when our chickens are free ranging, to see what they choose when given the choice, and I've identified a few types of weeds that they seem to particularly enjoy.

Grasses. An important part of any chicken's diet, grasses should be incorporated into your flock's daily menu whether or not they free range. Eggs laid by chickens who consume grass have a better hatch rate; the eggs have darker yolks and taste better. Grass pulls nutrients from the soil it grows in and can actually provide one-quarter of a hen's overall daily nutritional requirements. It fulfills ALL of her protein needs in the form of amino acids that are converted to protein. (Ryegrass, for example, has a 12–16% protein content.) Grass also provides Vitamins C and E, magnesium, iron, nitrogen and phosphorus.

Grass fulfills all protein needs.

The one downside to feeding grass is its fibrous consistency. The long, coarse and woody strands can get stuck in the crop, which may lead to sour crop or impacted crop, both serious conditions if not treated quickly. Normally, free range chickens won't suffer from crop issues since they snip off short pieces of the grass as they graze, so if you are cutting grass to feed to your chickens, be sure to cut it into very short (one inch or shorter) pieces for them and try to choose young, tender grass. And of course only feed grass that has not been chemically treated with pesticides or fertilizers.

Experiment with different foods and weeds to see what your hens like. Rule of thumb, if it's good for you, its good for them, with just a few exceptions. Chickens are true omnivores and will eat almost anything. Go heavier on the fruits, veggies, whole grains and lean meats, but remember that even healthy treats should only be fed in the afternoon once the chickens have had their fill of their regular feed. A balance of healthy treats can help cut down on feed costs while providing the chickens a variety of nutrients.

Here are some ideas for healthy treats:

Fruits

Watermelon is a great cool summer treat.

- **Seeded fruits,** including grapes, pears, apples (no seeds, see page 59: "Foods to Avoid"), grapes, papaya, pears and pomegranates
- **Pitted fruits,** including cherries, mangoes, peaches and plums
- **Melons,** including watermelon, honeydew and cantaloupe (cut in half; rind, flesh and seeds)
- **Berries,** including blackberries, blueberries, cranberries, elderberries, raspberries and strawberries (fruit, tops and leaves)
- **Dried fruits,** including blueberries, cranberries and raisins

Vegetables – Vegetables don't need to be cooked or even cut up, and in most cases can be fed in their entirety: flesh, skin, seeds, vines and leaves.

- **Gourds/Squash,** including cucumbers, pumpkins, summer and zucchini
- **Root vegetables,** including sweet potatoes, radishes, beets and carrots (all tops and leaves are fine)
- **Cruciferous,** including broccoli, cauliflower, Brussels sprouts, kale and cabbage
- **Greens,** including collards, chards, spinach and lettuce (limit iceberg, due to low nutritional value)
- **Other vegetables,** including bell peppers, corn, green beans and peas

Grains/Seeds/Nuts

- **Bread products,** including bread, cereals, crackers, grits grains, oats, cooked pasta and rice (all whole grain preferably, stale is fine)
- **Seeds,** including flax, millet, safflower and sunflower
- **Nuts,** including almonds, cashews, peanuts, walnuts (unsalted, shelled, chopped)
- **Popcorn,** air popped with no butter or salt

Meats/Protein

- **Insects/Bugs,** including crickets, earthworms, grubs, June bugs, mealworms, roaches and spiders (only those that have died a natural death or been stepped on; none that have been killed with bug spray!)
- **Meat,** including raw or cooked hamburger, cooked chicken, turkey, lamb and pork; raw or cooked steak scraps
- **Seafood,** including all types of fish and shellfish (cooked, including skin and shells)
- **Eggs,** fed raw or cooked

Flowers/Weeds (see the complete list of edible flowers on page 136)

- **Dandelion flowers/leaves**
- **Herbs** (see the complete list of edible herbs on page 136)
- **Marigolds**
- **Pansies**
- **Nasturtiums**
- **Violets**
- **Weeds** – none that have been sprayed with pesticides or other chemicals (see the complete list of edible weeds on page 136)

While also being a treat, herbs are essential for chickens' good health.

Foods to avoid

Certain foods contain substances that can be toxic to chickens or can cause other health issues. While most of these foods won't be immediately fatal to chickens, in large enough amounts they could lead to serious problems, or even eventually kill them. So I err on the side of caution and avoid them all, since there are plenty of other healthy choices for treats. The foods to avoid include:

- **Apple seeds** – The seeds contain trace amounts of cyanide.
- **Asparagus** – It can taint the taste of the eggs.
- **Avocado** – All parts of the avocado, flesh, pit, skin and leaves contain the toxin persin, which has been associated with myocardial necrosis in poultry.
- **Beans** – Dry, uncooked beans contain the natural insecticide, hemaglutin, which is toxic to chickens. But, soaking and cooking or sprouting the dried beans eliminates the toxin.
- **Caffeine** – Coffee, chocolate or tea bags should never be fed to chickens.
- **Citrus** – It is thought that an excess of citrus fruits (Vitamin C) reduces calcium absorption and can lead to thin-shelled eggs or even a drop in egg production.
- **Dairy/Yogurt** – Chickens are not able to digest the milk sugars in dairy products, which can cause diarrhea. So dairy products, while providing beneficial calcium, should be limited.
- **Eggplant** – A member of the nightshade family, eggplant should only be fed if completely mature and preferably cooked. Be aware that the leaves are toxic.
- **Onions** – Onions contain the substance theosulphate, which destroys red blood cells and can cause jaundice or anemia. (*Note:* Although garlic is in the onion family, it contains far less of the toxin and its health benefits outweigh any slight risk of anemia)
- **Potatoes** – All parts of white and red potatoes (potatoes are members of the nightshade family) contain the toxin solanine, which destroys red blood cells and can cause diarrhea and heart failure. Solanine is not completely destroyed by cooking

and should be avoided. Interestingly, sweet potatoes are part of the morning glory family (not a nightshade) and are perfectly safe.

- **Rhubarb** – This is another member of the nightshade family, with potentially toxic stalks and leaves.
- **Tomatoes** – Green tomatoes and tomato vines and leaves contain the same toxin (solanine) as potatoes. Once ripened, though, the solanine levels decrease to a minimal level. Our chickens love tomatoes but I only feed them very ripe fruit – and in moderation.

The question of goitrogens

There are several foods that contain goitrogens, substances that interfere with thyroid function. That might be of concern if a hen already is dealing with thyroid issues, but if fed in moderation shouldn't pose any problem. Their health benefits far outweigh any potential risk in most cases. These foods sometimes appear on lists of foods not to feed to chickens, so if you see any of these listed, that is why: broccoli, Brussels sprouts, cabbage, cauliflower, kale, millet, peaches, radishes, soybeans, spinach, tofu, and turnips. (Our chickens eat these foods on occasion, as do we, without issue.)

IT'S TREAT TIME!

Now that you know what kinds of foods can be fed as occasional treats, and which foods to stay away from, how about I show you a few ways to make treat time a bit more fun for the chickens? Many of these ideas are great boredom busters for winter when there isn't as much for the chickens to eat or they can't explore outside.

Berry box treat dispenser

Save your plastic berry boxes. They make excellent treat dispensers when filled with sunflower seeds, cracked corn or other grains. It won't take your chickens long to figure out that if they kick the boxes around the run the treats fall out.

Lettuce piñata

Hang a wire basket from the top of your coop or run, or repurpose an old shepherd's hook, and fill it with lettuce, cabbage, kale, broccoli or other greens for your chickens. This not only keeps the greens off the ground and out of the mud, the chickens will enjoy pecking at the greens as the basket swings back and forth.

Lettuce piñatas are a great way for chickens to get their greens.

Hanging apples

Another fun treat is cored apples filled with a mixture of peanut butter, honey, raisins and sunflower seeds – either hung from the side of the run or cut in half and handed out for the chickens to peck at. Be sure to provide plenty of water nearby with this sticky treat. As mentioned above, apple seeds contain trace amounts of cyanide, so should always be removed from apples before offering them to your flock.

Sunflower heads

I plant sunflowers for the chickens each spring. After they bloom, I hang the dried heads in the run or coop for nutritious treats through the winter.

Sprouts

You can sprout all kinds of seeds and beans for your chickens: clover, alfalfa, mung beans, peanuts, lentils, peas, quinoa, radish, mustard seeds, grains, clover, oats, garbanzo beans, sunflower and pumpkin seeds, among others. Check your local health food store or online for organic beans and seeds for sprouting. I use a glass canning jar with a piece of rubber shelf liner cut to fit across the top for my sprouts, which are first soaked and then rinsed several times a day and generally ready to eat after just a few days on a sunny windowsill.

Homemade seed blend
I mix my own seed blend for the chickens, consisting of millet, sunflower seeds, safflower seeds and dried mealworms. I store it in an empty Parmesan cheese or other shaker container for easy dispensing on the ground. But, when the ground is muddy, instead of tossing the seeds I use a hanging treat feeder. (See next page for how to make a hanging treat feeder.)

Summer salads are very nutritious for chickens.

Chicken salad
Here's a very nutritious, free treat that my chickens enjoy during the summer: dandelion greens, cut grass, parsley from the garden and wild berries. Drizzle a little apple cider vinegar/olive oil "dressing" over it for an added boost of nutrition.

Hanging Treat Feeder

If you have five dollars (or less, if you can repurpose things you already have) and about five minutes, you can make a handy hanging treat feeder for your chickens. Boredom can cause pecking, feather eating and other behavioral issues in your flock. Fill the feeder with greens, slices of watermelon, bread, homemade suet blocks or other treats and they will have a ball trying to grab a bite.

What you'll need:

2 6-inch pieces of 1x2 board
1 12-inch length of thin chain
A piece of ½-inch hardware cloth
    measuring 8x13 inches (any
    rough edges trimmed)
2 small eye hooks
Wire or string
Needle-nosed pliers
Staple gun/staples

What to do:

Fold the hardware cloth in half with the fold on the bottom and position one piece of wood at either side. Staple the hardware cloth to the wood, front and back. Attach an eyehook to each piece of wood at the open end (top) and then, using the needle-nose pliers, attach one end of the chain to each eyehook.

Hang your feeder in the run or coop using string or wire so it hangs about 6 to 8 inches off the ground. Fill with treats, stand back and watch the fun.

If you stick with mostly healthy treats and offer them in limited amounts, you will be providing your chickens with a well-rounded, balanced diet that will keep them happy and healthy.

Calcium-Rich Veggies for Strong Eggshells

Many vegetables are excellent natural sources of calcium. These top the list: broccoli, bok choy, Brussels sprouts, butternut squash, cabbage, celery, chard, collards, comfrey, dandelion greens, green beans, kale, mustard greens, raspberry leaves, red clover, turnip greens and watercress.

EIGHT

IN THE SUMMER

Chickens fare best in temperatures somewhere between 55 and 75 degrees Fahrenheit. Any higher than that and the heat can start to put unnatural stress on their bodies. While they will slowly acclimate to rising temperatures in the summer, a sudden heat wave is extremely hard on them and anything you can do to alleviate their discomfort can sometimes literally be the difference between life and death for your hens.

When it's hot

Chicken have a far easier time in cooler temperatures than warm ones. That's because they have a very hard time cooling themselves. They don't have sweat glands and instead rely on their respiratory system to help them cool off. They will pant and hold their wings away from their bodies to allow cool air to flow close to their bodies. They use their combs and wattles as radiators of sorts to release body heat. In warm weather, blood automatically flows to a chicken's comb and wattles, drawing it away from their vital organs, which over time takes a toll on their bodies.

Breeds with large combs, such as Andalusians, handle heat far better than chickens with smaller combs. Lighter colored breeds seem to handle the heat better than the darker breeds and smaller-bodied chickens, such as bantams and silkies, do better than heavier breeds. The Mediterranean breeds, including Andalusians, Egyptian Fayoumis,

Penedesencas and Sicilian Buttercups logically handle heat very well. Some other breeds that do well in the heat are Anconas, Campines, Hamburgs, Lakenvelders, Leghorns and Naked Necks.

It's very important to do all you can during the summer to keep your chickens cool. The effects of the heat on their bodies is cumulative, so by the end of the summer, they can be quite vulnerable to heat stroke; this can be fatal, especially in older hens, who have a harder time in the heat than do younger members of a flock.

When it's hot, chickens should be allowed to relax and not be unduly stressed. Children should be taught not to chase them or try to catch them, and your dog shouldn't be allowed to run the fence and bark at them – any time really, but most importantly when it's extremely hot.

Breeds with large combs fare better in warm climates.

Plenty of shade and fresh, cool water are the easiest, most basic ways to help your hens handle the heat. They will drink two to four times more water in the summer than in the winter. A great deal of water is needed to produce an egg, so access to clean, cool water is critical. Being deprived of a fresh water source for even a few hours can negatively affect egg production, resulting in fewer and smaller eggs with thinner shells.

Water

Given a choice, chickens prefer their water to be around 55 degrees, but will readily drink water that is much cooler than that. However, once their water gets too much warmer, they will stop drinking and can die of dehydration before they will drink it. Chickens generally won't even touch water that registers above 90 degrees, so when the air temperatures approach double digits, it's time for ice cubes or ice blocks in your chickens' water. Vitamins & Electrolytes or plain Pedialyte added to the water in extreme heat can help your chickens cope better with the heat, as can adding regular baking soda to the water in a 2% ratio.

I make my own electrolytes to add to my chickens' drinking water in the summer. Using common kitchen ingredients, it's great to mix up in a pinch when there is an impending heat wave. Replacing the electrolytes lost during times of oppressive heat could mean the difference in your chickens' chances of surviving.

Homemade Electrolytes

What you'll need:

1 cup water
2 teaspoons sugar
1/8 teaspoon salt
1/8 teaspoon baking soda

What to do:

Use full strength on severely ailing chickens; otherwise, mix into their drinking water as needed, one cup per gallon of water.

Cooling their heads and feet. Providing your chickens with deep tubs of water to stand in and in which to dunk their heads is a great way to help them cool down. Adding ice cubes or frozen water bottles to their drinking water is an even better idea.

Feeding Tips for Hot Weather

In the summer, try to feed your flock as early in the morning as you can, and then leave feed for them again just before dark; this way they aren't forced to eat in the middle of the day when it's hottest. Or rig up a low-watt light bulb in the coop and feed them overnight. If you do, be sure to leave water in the coop as well.

Nighttime heat in the coop

A big problem in the summer can be cooling down the coop at night when the temperatures hover around 100 degrees during the day and barely cool down from that through the night. The chickens will be reluctant to go to roost in the coop if it is too warm, so it's crucial to provide adequate ventilation, especially in the southern climates. As many vents and windows as you can manage will increase airflow and comfort (see Chapter 1, "In the Coop").

Can I leave my chickens outside in the run? When it's hot in the coop it might be tempting, but not only will it interfere with their routine (and chickens are very routine-oriented), it's not safe unless your run is 100% predator-proof: totally enclosed (top too) in chain link or welded wire fencing – sunk into the ground about a foot to deter digging predators, like raccoons foxes, coyotes, dogs and wolves. Remember that predators such as weasels, snakes and rats, can squeeze through very small spaces in fencing, so wrapping chicken wire around the lower few feet at a minimum is also recommended.

Cooling down the coop. On hot summer nights, I freeze water in gallon water or milk jugs, hang them from the roosts, and then set up an oscillating fan blowing on the jugs. It is amazing how much that will help to cool the air inside the coop and make for much more comfortable sleeping. I also lay some jugs on the floor of the coop so the chickens can perch on or next to them if they want to. The condensation from the melting ice creates nice, cool air and the fan helps circulate it.

Shade outside

In the run, shady areas filled with loose dirt allow the chickens to wriggle down into the cooler layer of dirt underneath to try and cool down. You can set up nesting boxes outdoors in the shade to provide cooler places for your chickens to lay their eggs than in a hot coop.

First Aid for Heat Exhaustion

Classic signs of heat exhaustion include excessive panting, holding the wings out (although most chickens will do this to some extent and it's perfectly normal – and beneficial – in high heat), a very pale comb and wattles, standing with their eyes closed, or maybe even lying down.

If you have a hen that seems to be suffering heat exhaustion or dehydration, get her somewhere cool as quickly as possible. Dunk her comb and soak her legs and feet in a tub of cool water to bring her body temperature down. Give her cold water to drink and some plain Pedialyte or homemade electrolytes (even Gatorade in a pinch), for added electrolytes and nutrients to replace what she has lost.

Treats for warm weather: some dos and don'ts

Summer treats differ a bit from winter treats. **Scratch** (cracked grains, including corn) should *not* be fed in the summer, as it can raise body temperatures. Excellent warm weather treats include **frozen watermelon, halved grapes** or **cantaloupe**, all of which will help hydrate in addition to cooling the chickens off. Other cooling treats: **frozen strawberries, blueberries, cucumber slices, bananas** (try rolling the banana in honey and chopped nuts and then freezing), and **peas**.

Mint tea. Mint naturally lowers body temperatures of both humans and animals, which is part of the reason mint juleps and mojitos are so popular in warmer climates as summer drinks. Using that same principle, I make it a point to brew mint tea for my chickens

when the temperatures rise. I just boil some fresh mint leaves in water, then let them steep and cool to room temperature. Once cool, I strain out the leaves (you can also choose to leave them in and let the chickens eat them), refrigerate the tea and serve it as desired, once it's chilled. Throw in some ice cubes and you'll help your flock cool off even more on hot summer days.

Mint ice pops. I like to make these for my chickens in the summer. I get out some ice cube trays or small freezer containers and make up a batch with **blueberries, strawberries** (fruit and tops), cut-up **watermelon, peas** and other fruits and vegetables, plus fresh **mint** (or **lemon balm**, also in the mint family) – along with water – and freeze it. Sometimes

Mint and fruit ice "pops" make a nice cooling treat for your flock.

70

I freeze a string in each "pop" so I can hang them from the run fencing, other times I just serve them up in a shallow dish. My chickens love pecking at the ice to get to the fruits frozen inside.

Confetti ice wreaths. I also freeze fruits and vegetables such as **corn, peas, carrots, blueberries, cranberries** and **grapes** in water in Bundt pans to make confetti ice wreaths. I tie them with a ribbon and hang them in the run in the shade. Or sometimes I'll serve the wreaths to my chickens in a shallow dish.

Freezing your vegetable cooking water. This is another nutritious and easy summer treat. Just save your vegetable cooking water and chill or freeze it. The chickens will not only love a change from the plain water but will also benefit from the nutrients leached out of the vegetables.

Your chickens will have a great start toward beating the heat if you can provide them with shade, good air flow, clean, cool water and frozen treats. It's so important to keep your chickens cool in summer. Anything you can think of to help them stay cool may not only save their lives, but will result in more consistent egg production through the summer – and will definitely be appreciated by your chickens.

HERBS TO USE IN THE SUMMER

Lemon balm	Mint

NINE

WHEN THEY'RE MOLTING

The heat of the summer has finally passed and the shorter days signal to the chickens that it's time to shed their old, broken, dirty feathers and grow new ones for the winter. This is known as a molt. The molt happens not only for aesthetic reasons but also for health reasons. Brand new feathers help trap warm air during the cold winter months better than old feathers. Any parasite eggs laid on the old feathers will be shed as well.

The first adult molt (after the chicks have lost their baby fluff and grown their "big girl" feathers) usually happens in the fall when chickens are about 18 months old. Thereafter, they will molt annually. A chick goes through several molts from hatch until it starts laying, usually at around five to six months. Generally, between one and six weeks after hatching, the chick will be growing its first feathers to replace the baby fluff. Then a series of partial molts takes place, during which larger and stiffer feathers grow in. The first is between seven and nine weeks, the next between 12 and 16 weeks, and the last between 20 and 22 weeks – at which point the pullet has finally grown all of its adult feathers and should begin laying shortly.

There are no hard and fast "rules" when it comes to molting. Some hens molt very quickly, completing the entire cycle in just a few weeks. Others take months. A molt will hardly be noticeable in some hens, while others will look like they have been through the spin cycle of your dryer. Nothing is sadder looking than a hen going through a rough molt!

The "harder" a chicken molts, most likely the better layer she is.

Your chickens will not only look bedraggled, they will most likely lose weight as well. Combs and wattles will shrink and be pale. Molting can really take a toll on a chickens' bodies, yet another reason to be sure they are in tip-top shape going into the fall.

Some chickens molt nearly continuously, at a very slow pace, just a few feathers at a time, so it's not even noticeable. Ironically, your best looking hens, with beautiful, glossy feathers year-round, are probably your worst layers!

Egg-laying during molting

Hens usually won't lay any eggs while they are molting. All of their energy and nutrients are funneled toward growing the new feathers (which are primarily protein), rather than creating eggs (which would compete for protein). On occasion, your better layers might continue laying partway into the molting period while just their head and neck are regrowing feathers, but they will stop once their wing feathers start to drop. They will not resume laying until they are completely done molting.

Better layers tend to molt very quickly (taking only 12 weeks or so), and shed feathers in huge chunks. Then they get right back to laying eggs. Poor layers aren't in any hurry to get back to the business of laying eggs and tend to draw the molt out longer, sometimes stretching it out for six months or more. It might not even be noticeable that they are molting, as they tend to shed only a few feathers at a time. *Note:* Roosters also molt and are infertile while they are molting. It is important to keep a rooster's body weight consistent during the molt or he can be left sterile if he loses too much weight.

What can trigger a molt

Although fewer hours of daylight are what most often trigger a molt, other stresses such as extreme heat, wide temperature swings, overcrowding, sickness, the presence of predators (or an actual predator attack), poor nutrition, lack of drinkable water or sudden shock can cause a hen to start molting at any time of the year.

Adding supplemental light in the coop (to keep egg production up through the winter) and then changing your mind and deciding to turn off the light midway through can also trigger a molt. That is because the chickens view the sudden reduction in daylight hours as an indication that fall is finally arriving – even though it's the middle of winter and an inopportune time to be half naked.

Choosing to light your coop to maintain production at a steady level even while the days grow shorter can cause your chickens to hold off on molting until winter. The artificial light throws off their body cycle and if they don't molt in the fall, as their normal biological clock dictates, eventually their body will tell them to shed feathers anyway. This is one reason I don't recommend artificial light.

Late summer to early fall, you will begin to notice feathers all over your chicken yard. You will start doing head counts to make sure no one is missing, perhaps the hapless victim of a predator attack. You will worry that they have all suddenly become cannibals, plucking each other's feathers out with abandon. But no worries, it's all perfectly normal.

You might notice your normally affectionate chickens shy away from being touched or held. The partially grown-in pin feathers are extremely sensitive to the touch and can bleed quite easily if damaged, as they contain blood-filled veins, unlike fully developed feathers. Some of your chickens may even appear to be not feeling well, hiding behind bushes or standing hunched over. While they are molting, it's almost as if they know they aren't looking their best!

The pattern of the molt

The molt always occurs in the same pattern. It starts at the head and neck, then works its way down the saddle and back, breast, abdomen, wings and then to the tail. The new feather shafts literally push the old quills out, so there aren't generally huge bare patches; you will already start to see a new pin feather beginning to grow as soon as the old one falls out. As each new shaft pushes through the skin, it is covered in a waxy coating that which will eventually fall off as the chicken preens, allowing the new feather to emerge and unfurl.

Molting always follows the same, very distinctive pattern.

If you do see large bare patches on your hens without the characteristic molt pattern, that generally means there is a pecking or feather-eating issue in your flock. Bare patches around the vent can signal an external parasite infestation, and a bare breast usually means you've got a broody hen. These conditions are all discussed later on in the book.

When your chickens are going through a molt, try to keep them calm and on a routine – no kids running around and trying to pick them up, please.

Attention to their feed during molt

Access to plenty of good-quality feed as well as extra nutrition is extremely beneficial. During this time of extreme stress on their bodies, it becomes even more important to add **apple cider vinegar** to their water (see recipe on page 52). Certain herbs can help with feather regrowth, especially some of the aromatic herbs. These include **anise, dill, fennel, garlic, mint** and **parsley**. Fresh herbs probably won't be in season while your chickens are molting, but dried, crushed herbs can be easily mixed into their feed. Another good addition is mineral-rich **dried seaweed** or **kelp**.

The rate of feather loss and regrowth varies from hen to hen.

Extra protein. Increased levels of protein become critical because, besides needing the protein to grow strong new feathers, chickens whose diets are deficient in protein tend to start pulling and eating each others' feathers (chickens are drawn to blood and the color red anyway, and the growing feather shafts are full of blood). This can spell disaster if protein levels in their food intake are not kept high. Feathers are comprised of approximately 80 to 90% protein, 8% water and 1% water-insoluble fats, so providing your molting chickens with extra protein during their molt is critical, as is constant access to fresh, clean drinking water.

Good protein sources include **bugs, beetles, earthworms, mealworms, black oil sunflower seeds, scrambled eggs, green beans, peas, cod liver oil, mackerel, sardines, canned tuna, raw or cooked chopped hamburger** and **meat scraps**. Scratch and cracked corn should be limited during the molt since neither provides very much nutrition.

Protein-rich "Molt Muffins"

I have created an easy recipe for a protein-rich treat that I feed to my hens while they are working hard to grow back their feathers. Molt Muffins are easy to make and the chickens love them. The peanut butter, sunflower seeds, oats, wheat germ and mealworms provide much-needed protein, while the powdered milk and molasses provide additional calcium, both of which help the molt go faster and more smoothly. *Note:* always provide clean, fresh water when feeding anything to your flock, but especially when peanut butter is involved!

Molt Muffins

Makes 6 muffins

What you'll need:
½ cup old-fashioned oats
½ cup shelled sunflower seeds
½ cup dried mealworms
¼ cup wheat germ
2 tablespoons powdered milk
¼ cup raisins
¼ cup coconut oil, warmed to liquid consistency
1 tablespoon blackstrap molasses
1 cup natural unsalted peanut butter

To hang your 'muffins,' you will also need six large buttons (1" in diameter or larger so the chickens can't swallow them) and baker's twine.

What to do:

In a large bowl, combine the dry ingredients. Stir in the coconut oil and molasses, then add the peanut butter and mix well. Set aside.

Place six paper cupcake liners in a muffin tin. Thread a long length of baker's twine through two holes in each of the buttonholes and place one button in the center of each cup, leaving the ends of the twine hanging over the side. Spoon muffin mix evenly into the cups, making sure your twine/button is centered in each cup. Refrigerate the muffins until firm, remove them from the paper liners, hang them in your run and watch your hens enjoy their nutritious treat.

HERBS TO FEED WHILE THEY'RE MOLTING

Anise	Fennel	Mint
Dill	Garlic	Parsley

TEN

IN THE WINTER

Winter is approaching, and your chickens are ready to brave the elements with their brand new feathers. In the winter, they will fluff their feathers to trap warm air next to their bodies to help insulate themselves from the cold air. Chickens handle cold far better than heat, but that doesn't mean they won't appreciate a few creature comforts on those cold, blustery days. The most important thing is that your coop is dry and draft-free, with good cross airflow and ventilation that is higher up than the roosts. Inadequate ventilation will lead to high moisture levels, which can contribute to both frostbite and respiratory issues. Frostbite is partially caused by damp conditions, so having good ventilation not only provides clean fresh air but helps prevent frostbite.

Roosts should be wide enough that your hens' feet are flat when they roost and completely covered by their bodies from the top and the roosting board from underneath. This helps to prevent frostbite on their toes.

Cold-hardy breeds

Chickens with larger bodies and smaller combs are more cold-hardy than smaller breeds with large combs. Some particularly cold-hardy breeds include Ameraucanas, Australorps, Barnevelders, Barred Plymouth Rocks, Brahmas, Buckeyes, Buff Orpingtons, Cochins,

Chickens with small combs handle cold better than those with larger combs.

Dominiques, New Hampshire Reds, Rhode Island Reds, Sussex, Welsummers and Wyandottes.

Although conventional wisdom says to apply a layer of ointment or salve (coconut oil, petroleum jelly, etc.) to your chickens' larger combs on particularly cold nights, that is far easier said than actually done. Your chicken will hate it and immediately rub her coconut-oil-slathered comb on the ground, resulting in a dirt-covered, coconut-oil-smeared comb. I say skip it and instead choose cold-hardy breeds, if you live in an exceptionally cold climate, and let your hens tuck their heads under their wing to sleep.

Winter comfort in the coop

Heat should not be necessary in your coop unless you have very young chickens, breeds that are not cold-hardy, or injured or sick chickens. Heating your coop doesn't allow the chickens' bodies to gradually adjust to the colder temperatures, and a power loss or other interruption in the heat flow could kill them, since they would not be used to the cold. You also run the risk of a coop fire if you use a heat lamp or other heat source there. Dry bedding plus a heat source, electricity and flighty chickens is not a good mix.

A better way to generate heat in your coop is to use the Deep Litter Method. It is an old-timers' trick that allows manure and bedding to accumulate and decompose inside the coop all winter. Then in the spring, you clean the whole thing out and have beautiful compost for your spring garden.

The first few winters we raised chickens, I would trudge down in the snow and ice to clean out the coop every other week or so. I would remove all the straw bedding and replace it with new straw. The old soiled bedding would sit, partially frozen, in our compost pile until spring. I didn't enjoy doing it; it didn't seem practical and I knew there had to be a better way. So I tried the old-timers' way – with deep litter.

The Deep Litter Method

This basically consists of turning over the soiled bedding, adding a new layer, and allowing the chicken droppings to decompose on the floor of the coop all winter. The decomposing process will create heat, keeping your coop warm naturally. As a further bonus, as in composting, beneficial microbes will start to grow. These microbes help control pathogens and prevent parasite eggs from developing, making your

Natural composting using the Deep Litter Method.

chickens less susceptible to diseases such as coccidiosis or mite infestations. Then in spring, all you do is just clean the whole thing out and dump it into your compost pile.

When using this method, you should use pine shavings as your bottom layer. Starting with the 6-inch layer of pine shavings on the floor, each morning I use a rake to turn over the litter so the soiled bedding from the night before ends up on the bottom. I continue doing that each day, adding straw (or more pine shavings) as needed to eventually build up to a 12-inch-deep layer. Nothing is removed, but rather turned over to expose new straw. (You can also use dry grass clippings, leaves, pine needles, or a combination.)

Chicken manure is very high in nitrogen. Mixing it with a source of carbon (either straw, shavings or dry leaves) will balance the mixture and hasten the rate of decomposition. It is important that your composting material contain oxygen, which means turning is crucial. Fortunately, the hens will help you with that part, especially if you get in the habit of tossing some scratch into the coop for them before bedtime. When they wake up in the morning, they will learn to sift through the litter to find the scratch you left the night before. The turning and introduction of oxygen will also reduce an unhealthy buildup of ammonia fumes. If done correctly, your coop shouldn't smell of ammonia or manure.

After just a few weeks, the droppings, shavings and straw will start to decompose and you will end up with a fine dirt on the bottom. Continue in this manner all winter. When spring arrives, you're ready to add great compost to your pile.

Some caveats before you start:

- Your coop must have good ventilation. If you smell even a hint of ammonia, you need to clean the entire coop out, put down a new layer of shavings and start over. Ammonia fumes can cause eye and sinus irritation in your flock. *Note:* Diatomaceous earth should not be used in conjunction with the Deep Litter Method; it will kill the good microbes and it isn't beneficial to have in your composted soil since it can kill "good" bugs.

- The Deep Litter Method is not appropriate during the warmer months since it does generate quite a bit of heat in the coop, which you only want in the winter.

Winter in the run

Chickens don't generally like to walk on snow. One way to entice them out on nice days is to put down some straw to make a path from the coop door to a sunny, sheltered spot of the run. Except on the most frigid days, it's best to leave the coop door open and let the chickens decide if they want to go outside or not. They'll be more likely to come out if you create a wind block in one corner of their run and set up some stumps, logs or branches (or even wooden pallets) as outdoor roosts to give them something to stand or perch on that's up off the cold ground.

Keeping their water from freezing

Even in the winter, it's important that your chickens drink enough water and have unlimited access to unfrozen water. I use an electric water bowl for dogs to keep their water thawed. I also set black rubber tubs in the sun with a few ping pong balls floating in them. Just the slightest breeze will keep the balls bobbing and the water unfrozen. Your chickens will appreciate being supplied with warm tap water once or twice a day as well.

Fresh air is very important in the winter, so let your chickens have the option to go outside if they wish.

Herbal tea and vegetable water

Although chickens generally don't like hot liquids, a bit of room temperature herbal tea is something they seem to enjoy on a cold winter day. It gives them a welcome, and nutritious, change from plain water. I dry herbs over the summer to use all winter in teas for my chickens. **Dried oregano, thyme, tarragon** and **basil** make nice tea for them. Just sprinkle the dried herbs into a saucepan full of water, boil and let steep for ten minutes, then cool to room temperature before serving.

I also save my vegetable cooking water when I am making dinner and serve that to our chickens. Any leftover vegetables and trimmings, along with leftover rice or pasta, mixed into the reserved cooking water makes a nice soup for the chickens the following morning. Leftover vegetable cooking water also can be poured over old-fashioned oats or even layer feed to make a nice, warm breakfast treat for your flock.

Egg production in winter

Just as the shorter days of fall signal to the chickens that it's time to molt, the short winter days will result in a drop in egg production for most breeds. A hen needs 14 to

Egg production typically drops in the winter season.

16 hours of daylight for the ovaries to be stimulated to release an egg. It is possible to keep egg production constant through the winter by adding artificial light in your coop each morning (to increase the total "daylight" hours to the proper range), but I let our chickens take the respite that nature intended. Coming out of a molt, their bodies are depleted of their protein and calcium stores. I like to give the chickens a break for the winter and let them take their time recovering so they are in tip-top shape come spring, when they will resume laying in earnest naturally.

While eggs are plentiful in the summer, I freeze the excess to use through the winter. This way, we still have a supply of eggs through the win-ter and the chickens have a chance to let their bodies replenish the nutrients lost during the molt. The defrosted eggs are wonderful for baking or scrambling. In fact, I don't even notice a difference between fresh eggs or the defrosted ones.

Freezing eggs is easy. Just lightly whisk them, trying not to incorporate too much air. Then whisk in 1/4 teaspoon of salt per cup of egg mixture, or 1 tablespoon of sugar, depending on what you plan on using the eggs for – add salt to eggs to be used for scrambling or sugar to eggs to be used for baking. This will help prevent the frozen eggs from becoming grainy. Next, measure out 3-tablespoon portions into ice cube trays and freeze. Each portion equals one egg. When they are frozen, pop them out and store them in the freezer in freezer bags labeled either "salted" or "sugared." When you want to use them, grab as many cubes/eggs as the recipe calls for and let them defrost overnight in the refrigerator. If you are scrambling them, you can add them right to your hot skillet. The defrosted eggs should only be used in recipes where they will be fully cooked. They will last up to 6 months in the freezer. You can also separate your eggs and freeze the yolks and whites separately. Separated whites don't need any salt or sugar added.

Feed and scratch for winter

Even if the chickens are taking a break from laying, they should still be fed their layer feed and also provided free-choice oyster or eggshell. Winter is also the time when scratch makes a great treat, especially just before bedtime. Scratch is considered a "treat," not a substitute for layer feed. It contains only about half the protein of feed, but the body heat generated digesting the corn and grains in the scratch helps keep chickens warm during cold nights. It also gives them something to do when there aren't any bugs or grass in the run to keep them occupied.

As an alternative to commercial scratch, you can easily mix up your own using bulk grains purchased from your feed store or local grocery store. I use a blend of cracked corn, oats, barley, wheat, flax seed, sunflower seed and raisins. The girls love it.

Cranberry Scratch Wreath

I make scratch wreaths for the chickens during the winter as well as summer. They are easy to make and will keep your chickens from getting cabin fever during long winter days when there isn't much to do outside.

What you'll need:

Cooking spray
Bundt pan
½ cup cool water
3 envelopes Knox unflavored gelatin
1½ cups boiling water
1 cup bacon, suet or hamburger grease, heated to liquefy
8 cups of a mixture of scratch, sunflower seeds, cracked corn, raisins, nuts or seeds
20 fresh or frozen cranberries
Ribbon

What to do:

Spray Bundt pan with cooking spray and set aside. In a measuring cup, dissolve the gelatin in the cool water and let sit for a minute. Pour the boiling water into a medium bowl and whisk in the gelatin to combine.

In a large mixing bowl, combine the seeds and nuts. Stir in the grease and then pour in the liquid gelatin. Mix well with a wooden spoon to be sure all the nuts and seeds are well coated and all the liquid is absorbed. Place the cranberries in rows in the indentations in the pan (I used three in half the indentations and two in every other indentation). Then carefully spoon the seed mixture into the pan. Press down with the spoon to pack it well.

Put the Bundt pan in the refrigerator overnight to set. The next day, take the wreath out of the refrigerator and let come to room temperature. Invert and tap gently on the countertop to unmold. Tie to your run fencing and let the chickens snack to their hearts' content.

Suet blocks

You can also make suet blocks for your chickens (see next page). Of course you can buy commercial suet blocks, but I prefer to make my own. That way, I know exactly what is going into them, and it's a great way to use the grease that would otherwise just be thrown away. The blocks are easy and very inexpensive to make if you save grease any time you cook burgers, bacon or other meats throughout the year. The added fat will help keep your chickens warm in the winter. It provides nearly twice the sustained energy of carbs and also slows the rate of food digestion, thereby increasing the absorption of the nutrients in the feed your flock eats. While I normally limit the amount of fat or grease I give to the chickens, I do save all our hamburger grease in the freezer until the winter so I can make some suet blocks for them.

Choosing to keep the process simple, I don't render the fat or otherwise make it stable for long-term or room temperature storage, so the suet should be kept in the freezer until ready to use and then only fed in portions that your chickens will eat fairly quickly.

Commercial suet blocks are available, but I prefer to make my own.

Homemade Suet Block

What you'll need:

Grease/fat (i.e., from cooking meatloaf, burgers, steak or low-salt bacon)

Unsalted nuts

Sunflower seeds

Cracked corn

Raisins

Cayenne pepper

Heat/freezer-safe container

Suet cage

What to do:

You can use any heat- and freezer-safe container you happen to have in which to store your suet. I use small, square casserole dishes because they make suet blocks the perfect size to fit into my feeders, but you can use any container you wish. You can even use a larger square cake pan and just cut the finished suet into smaller pieces to fit your suet cages.

Chop some assorted unsalted nuts (peanuts are especially nutritious and a good source of unsaturated fat). Arrange them in each dish and then sprinkle some raisins, seeds and cracked corn on top, along with a healthy shake of cayenne pepper. The cayenne helps to heat the body naturally and also stimulates the hens' circulatory system. This is extremely beneficial during the cold winter months. Then

when you cook meat, save and drain your grease (it's okay if there are a few random tidbits of meat in it). Let the grease cool just a bit, then pour careful- ly over the nut mixture. Stir to blend and then put the dish into the freezer. You can continue to add "layers" to your container each time you have leftover grease, adding more nuts, raisins and cayenne as needed. You can also add other dried fruits or seeds. Bacon fat can be used, but in moderation, due to the higher salt/nitrate content compared to other grease. I store the suet in the freezer until winter and then start doling it out on cold days.

To serve, remove the container from the freezer and run a butter knife along the outer edge. Turn the container upside down and gen- tly tap on the counter. Your suet block should pop right out. I put the block in a bird suet feeder, which is perfect for my homemade suet blocks. It keeps them out of the dirt and off the ground. I feed only what my hens will eat over the course of an afternoon or so, to be sure the suet doesn't go rancid (which really isn't much of an issue when temps go below freezing anyway!). But trust me, the suet won't last that long! I feel good knowing that I am not only helping our chickens deal with the cold, but also making good use of grease that I would otherwise throw out.

Edible garlands

Another fun, easy way to treat your chickens is to make them edible garlands. String **popcorn, grapes, cranberries, walnuts, Brussels sprouts, radishes,** or even **hard-boiled eggs** onto twine, then hang them in the run for the chickens to eat.

AVOIDING WINTER BOREDOM

Boredom can be a problem in the winter, when there aren't weeds and grass to munch on, bugs to eat, butterflies to chase or dirt to bathe in – and there's not much opportunity for them to sun themselves. Anything you can do to keep your chickens occupied can help prevent pecking and squabbling within the flock. Bored chickens are not happy chickens. Bored chickens tend to start pecking at each other and themselves, resulting in feather loss or worse. Once a flock sees blood, it can whip the chickens into a frenzy and they will sometimes actually kill flock-mates – purely out of boredom and pecking order clashes. Boredom pecking is more likely to happen if your coop and run are too small and the chickens don't have adequate space and can't spend as much time outside during the cold months. Boredom can also lead to egg eating, which can be a tough habit to break once it starts.

Chickens need their outdoor play time – even in winter.

Three fun chicken activities

- For their great entertainment, provide piles of straw, hay or leaves for your flock; and even better, toss in a handful of scratch or sunflower seeds.
- Hang a mirror in your coop or run. It will amuse your chickens more than you can imagine; although if you have a rooster, take care, because roosters will tend to attack the "newcomer."
- On nice days, even when there's snow on the ground, give your chickens a chance to bathe by providing them a temporary dust bath in a kiddie pool or other large container.

Doing these few small things for your chickens can help them get through the winter more easily – and will help them, and you, sleep better at night.

HERBS TO USE IN THE WINTER		
Thyme	Oregano	Basil

ELEVEN

WHEN THEY'RE BROODY

When a hen decides that she wants to sit on a "clutch" of eggs until they hatch, she is referred to as a broody hen, or as "going broody." One of your first signs that you have a broody hen on your hands will be seeing her in the same box every time you go into the coop. If you try and take eggs out from under her, she might peck at your hand, fluff up in indignation or growl at you. She also will start clucking a deep, throaty cluck (that is the "voice" she will eventually use to call her chicks).

Broody hens pluck out their breast feathers to literally "feather" their nests, providing a soft cushion for the eggs, and also so the eggs will have direct contact with the heat and moisture from their bare skin. Research has shown that wild birds will also line their nests with medicinal weeds, fresh herbs and flowers, especially those that contain essential oils. The herbs help repel parasites and keep flies away from the setting bird, but even more importantly, newly hatched baby birds benefit by rubbing against the herbs and weeds during the first few days of life. So why wouldn't the same apply to chickens? Going on the assumption that chicks hatched in the wild would hatch in a nest lined with various herbs and beneficial weeds, I add them to my nesting boxes for my laying hens and baby chicks to reap the same benefits.

Adding herbs to your broody hen's nest can have multiple benefits, not only for the hen but for the chicks that will eventually hatch. Herbs impart protection from parasites and pathogens, help strengthen their immune and other systems, and offer other health benefits as they peck at them and also eat some.

I believe that the herbs can help calm and encourage broody hens. Studies show that a hen's stress level goes up when she is searching for a place to lay her eggs; she wants a place that she feels will be safe and protected, so she won't be discovered by predators and the eggs will have a good chance of hatching. The studies reveal that the most relaxed hens are those who find dark, secluded places to lay – and also, that relaxed hens tend to sit longer on the eggs. Oftentimes, a broody will lose interest and stop sitting on fertile eggs before the hatch date, so ensuring that your broody feels safe and calm will increase the chance that she will sit for the duration.

This broody hen has plucked some of her breast feathers to "feather" her nest.

Ever since adding the herbs to our nesting boxes, I have seen a marked increase in the frequency of our hens going broody. I have never read any studies on the topic of using herbs to calm hens, but from my own observations, I believe that the herbs do relax and encourage the broody hens. I also wonder if the herbs I have been adding to the nesting boxes are a signal to the hens that the nest is optimal for hatching healthy chicks, and that has been one of the reasons for more broody hens than usual.

Calming herbs for the nesting box

Lavender, bee balm and **yarrow** are all calming herbs and perfect choices for your nesting box when you have a broody sitting. But lavender is the herb I would use first and foremost to encourage and assist a broody. Besides its calming and aromatic properties, lavender increases blood circulation, which is very important for hens sitting on eggs; they

don't often leave the nest to stretch their legs or get much exercise. Lavender is beneficial to keeping their blood flowing, as are **cayenne pepper, clover, violets** and **garlic** (garlic also helps immune systems and will make for a more healthy mother hen).

Keeping parasites away

Broody hens have a tendency to contract mites and other parasites since they don't get out to dust bathe often, if at all, and the warm, dark space underneath them is a prime breed- ing ground for parasites. Herbs can safely and naturally prevent mites from taking up resi-

Herbs in your nesting boxes can provide all sorts of benefits to sitting hens.

dence, specifically, **mint, catnip** and **thyme** help repel insects and parasites. **Lemon balm** can help prevent leg mites in setting hens.

Fragrant snacks

Pretty and naturally aromatic, **rose petals** can help combat some of the odor of the brood- ies, while also acting as a nutritious snack for your setting hens. The hen will only leave the nest a few times a day to eat and drink and will appreciate something to munch on in between trips out to the yard.

Herbs for a healthy nesting box

Adding a few **bay leaves, echinacea, dandelion greens** and **garlic cloves** to your nesting boxes can help boost the immune systems of your hens, and eventually the newly hatched baby chicks, when they rub against them or nibble on them. **Bay leaves, lemongrass** and **rosemary** will also help repel pesky, disease-ridden flies.

A bit of **oregano** might help combat coccidiosis, which is found in nearly every chicken coop and run in the world. Hens build up a tolerance and eventually an immunity by being exposed to small amounts of the coccidia pathogen, but newly hatched chicks greatly benefit from a bit of help in the beginning against the coccidia pathogens they invariably will come in contact with. Instead of using medicated feed, I offer oregano to my baby chicks on a regular basis and also add it to my nesting boxes.

Sitting and incubating

While your hen is sitting on her eggs, be sure to provide feed and water close by. Other than that, she knows what to do and you don't need to do a thing; although it might be a good idea to mark the eggs she is incubating with a pencil, just in case other chickens sneak in and lay more eggs in her nest while she's on one of her short breaks. That way, you will know to remove the new eggs and won't have a perpetual nest of eggs hatching in succession!

Make sure there is a nice, thick layer of nesting box bedding to keep the eggs from breaking. Your broody hen will turn the eggs periodically and also move those in the outer ring closer to the center, giving all the eggs equal time in the middle; that way, they all stay uniformly warmed. She will usually kick out any infertile or non-developing eggs. She will take short breaks to eat and drink, eliminate waste and stretch her legs, but for the most part of the entire 21 days, she will sit on those eggs and not do much else.

Now you've got chicks!

When the chicks hatch, they will need to be in a place where they won't fall to the ground or be in any danger. You can move them, mother hen and all, to a pen or crate where they will be safe from predators and the other flock members, who aren't always welcoming to new chicks. (It's best not to try moving the hen and eggs once she has started sitting, because that can cause her to stop. But if she has chosen a location that is not protected from the elements or predators, or is otherwise unsafe, you should try and carefully relocate the entire nest to somewhere safer.)

Chick starter feed should be provided in a small dish (it's fine for the hen to eat that while she is mothering the chicks), as well as a shallow dish of water that the chicks won't fall into and drown. The hen will show the chicks how to find food and drink and watch over them until she feels they are ready to be on their own. At that point, they all should be ready to join the flock.

When a broody won't stop sitting

While broodies can be a godsend if you are interested in hatching chicks from fertile eggs, hens will also try to sit on and hatch infertile eggs. It

Make sure the hen has a thick layer of bedding to keep the eggs from breaking.

isn't in her (or your) best interest to let a hen sit on eggs that will never hatch or on an empty nest. When she is sitting on fertile eggs and the hatching chicks start to peep and pop out of their shells, that's the signal to her body that the broody period is over. If eggs never hatch under her, she won't get that signal and sometimes will just continue to sit far past the normal 21-day incubation period, which can be detrimental to her health and well-being.

Since a setting hen only takes short breaks to eat and drink a bit and stretch her legs, most broodies lose a substantial amount of weight from reduced feed intake. In addition, broodies take up valuable nesting boxes that your other laying hens want, and jockeying for position can often lead to broken eggs. Broody hens don't lay eggs during the time they are broody (as well as for several weeks or months afterwards, depending on how long they were sitting), and it is thought that broodiness is "catching" to a certain extent, so leaving one hen sitting can often induce other hens to go broody as well. Your production levels can be greatly affected by letting a broody just continue to sit.

What you can do

It is far healthier for you to break a broody's sitting spell and get her back outside with the others – scratching for bugs, running around and taking dust baths – as quickly as possible. There are a couple of ways to do this, and it all depends on how stubborn your hen is and how strong her broody trait.

Breaking a broody hen requires you to be more stubborn than she is.

Some hens will give up on their own if you just keep removing the eggs from under them and taking them out of the nest as many times a day as you can. Others will stop sitting if you block the particular box they have chosen, refusing to just move to another box. Or you can try removing all the nesting material in the box in which the broody is sitting, or putting a frozen water bottle or ice pack under her. Other broodies are more stubborn and require actually being shut out of the coop or denied access to all of the nesting boxes. Once the other hens have finished laying their eggs for the day, just shut the coop door to force your broody to spend time outside.

If all else fails

As a last resort, you can pen your offender in an open cage such as a dog crate inside the run, preferably elevated and without any nesting material. This serves to cool her abdomen and vent area, which can help break her broodiness. Return her to the flock in a few

days. If she heads right back to the nesting boxes, it's back into the crate for another day or so. I have never had to take it that far. I find that gently removing the broody hen from the nest, taking any eggs she is sitting on, and then releasing her at the far end of the run where I have some special treats for everyone, generally works in just a few days. (Personally, I think some broodies are happy to give up and are just looking for an excuse to stop. After all, what's the fun of sitting alone in a dark box in the coop when you could be outside playing and being fed yummy treats?)

Whether you are trying to encourage a broody to sit on fertile eggs and hatch chicks or not, adding some herbs to the nesting boxes can help make your chickens' time in the box more healthful, relaxing and beneficial to them.

After she lays an egg, a hen will cackle and sing her "egg song." Some believe that she is announcing to the other hens that she has laid an egg, inviting them to come lay their eggs in the same nest, to contribute to a community "clutch." Others believe she is drawing attention away from the egg and toward herself, to keep her egg safe from predators. Either way, the hens are definitely trying to communicate something!

HERBS TO USE WHEN THEY'RE BROODY

Bay leaves	Garlic	Rose petals
Bee balm	Lavender	Rosemary
Catnip	Lemon balm	Thyme
Cloves	Lemongrass (citronella)	Yarrow
Dandelion greens	Mint	
Echinacea	Oregano	

TWELVE

IN THE BROODER

Our first chicks came from the local feed store. I approached my foray into chicken keeping as if I were in a French bakery, completely overwhelmed by all the delectable choices and not being able to choose between them, finally deciding, "I'll just take two of each." We came home with a nice mix of Wyandottes, Buff Orpingtons and Rhode Island Reds. In the ensuing years, I have bought chicks on Craigslist, from hatcheries, breeders and hatched my own. But no matter how you end up with your chicks, they all need a safe, warm place in which to grow.

Once you've done your research and know how many chickens you are allowed in your area, then figure out which breeds will be best for your family, and decide whether to buy your chicks locally or via mail order. As soon as you have a delivery date set, you will need to think about setting up a brooder with heat, feed and water. This will be their home for the first several weeks of their lives.

Here's what you will need before your chicks arrive:

- **Brooder** – either homemade or store-bought (i.e., cardboard box, plastic storage tote, dog crate)
- **Rubber shelf liner/newspaper**
- **Pine shavings**
- **Wooden dowels**

- **Heat lamp** with spare bulb, or Brinsea EcoGlow
- **Chick starter feed**
- **Chick-sized grit**
- **Sugar water** or plain Pedialyte
- **Chick-sized waterer** with marbles or stones
- **Chick-sized feeder**

Chicks need special-sized feeders.

The brooder area. Find a draft-free, quiet location in your garage or house – such as a spare bedroom, bathroom or mud room – safe from curious cats, dogs and children.

Chicks need to be kept warm. Before they grow in their feathers, they don't have much luck self-regulating their body temperature, so your brooder needs to be heated and free of drafts. Because it also needs to be protected from family pets and children, a covered brooder is optimal. As the chicks get older, the cover will keep them inside, since baby chicks can easily flap, flutter and fly out of boxes and totes.

A plastic storage tote makes an ideal brooder for chicks and is easy to make.

Making your brooder

A large plastic storage tote with the top cut out and replaced with ½-inch hardware cloth becomes an easy-to-make brooder that will keep your chicks safe. The solid sides prevent drafts and keep in the heat.

Slip-proof the brooder floor. To prevent the chicks slipping on the slick bottom of your brooder and possibly risking spraddle leg – which causes a chick's legs to splay out to the sides, making it difficult to walk – I line my brooder with several layers of **newspaper** and then add a sheet of **rubber shelf liner** on top. Newspaper alone is too slippery, especially when it gets wet, but it's a wonderful absorbent layer under the shelf liner.

After a few days, once the chicks learn what is food and what is not, a thin layer of **pine shavings** can be added. The shavings should be the larger "chip" variety, not sawdust, which is too easily eaten and also too dusty. The brooder will need to be emptied of the soiled shavings and newspaper as needed, rinsed with white vinegar/water and left to dry before new shavings are added and the chicks put back in. The shelf liner is easily removed, rinsed off and reused. A cardboard box can serve as a temporary brooder for the chicks while their tote is being cleaned.

Sand: not recommended.
While you might be tempted to use sand and just scoop

I add pine shavings on top of a rubber shelf liner once the chicks know what their food is.

out the brooder like you would a kitty litter box, that is not recommended by many poultry experts. The chicks will be tempted to eat the sand and can end up with impacted crops, which can kill them. The sodden sand collects in their crop and prevents them from not only getting adequate nutrition from their feed (since they will stop eating), but can also press against their windpipe and suffocate them. Also, when they poop in sand, their feces will become "breaded" with sand and it's possible for the chicks to then see it as food and eat it, since their instinct is to eat what is at their feet. Sand can harbor E. coli – extremely detrimental to your chicks. Sand is often made up of silica particles; the particles can be inhaled and lead to a respiratory illness called silicosis later in life. Additionally, in the brooder, sand can get very hot under the heat lamp for little feet to stand on. For those reasons, pine shavings are recommended.

> **WARNING:** never use cedar shavings, which can be toxic.
> *(see page 105 for more about bedding and which materials to avoid)*

Perches. I have wooden dowels screwed into the sides of my brooder for the chicks to practice balancing and roosting on. At just a few days old, they will already start perching on them, and I believe it helps make their transition to a coop and sleeping on a wooden roost more fluid.

Heating the brooder

A well-secured red heat lamp positioned so one end of your brooder stays a constant temperature is necessary to keeping the chicks warm when they first arrive. A red light helps prevent feather picking and pecking issues and stresses chicks less than a white light will. I normally use a white light by day, and then switch to a red light at night, to allow the chicks a chance to rest and relax under the red light.

Rule of Thumb Temperature Chart

1st week 95° F	4th week 80° F
2nd week 90° F	5th week 75° F
3rd week 85° F	6th week 70° F

(At six weeks old, if the daytime temperatures are close to where your brooder temperature is, the chicks can start spending time outside, at least during the day when it's warm and sunny.)

Wooden dowels make great perches.

The temperature in the box should be 95 degrees the first week, and then reduced by 5 degrees per week. I have a thermometer attached to one wall of the brooder so I can regulate the temperature, which I do by adjusting the height of the light.

While the temperature chart is a handy reference, the best way to judge the temperature in your brooder is to watch the chicks' behavior. If they are cold, they will be huddled under the light, peeping loudly. If they are too hot, they might be holding out their wings from their bodies or panting, and clustered as far from the heat as possible. Happy, well-heated chicks will be scampering around the brooder, cheeping contentedly.

By positioning the heat at one end of the brooder, you allow the chicks to decide where they feel most comfortable. I tend to leave the water and feed somewhere in the middle, not directly under the light.

Be sure you have a spare bulb...just in case. You don't want your bulb burning out, necessitating an emergency run out to the feed store to try and find a new one in a hurry. It's very important to keep the brooder box temperature constant. Chicks are extremely susceptible to being chilled before they are fully feathered.

Now that you have your brooder area set up, you're ready to add some chicks!

When you bring them home

Once you get your chicks home, check each one for "pasty butt" and clean off any poop that has accumulated on or around their vent with a Q-tip that is moistened with warm water or vegetable oil. Pasty butt literally stops up the chick so it can't excrete; it can be potentially fatal in a very short period of time. Caused by stress or extreme temperature changes like those often endured during travel from the hatchery, it is not uncommon in chicks that have been shipped. Keep checking every day until all the chicks seem free of any problems. A bit of **raw oats** or **corn meal** added to their feed can help prevent or clear up pasty butt, as can a sprinkle of **probiotic powder** on top of their feed.

As you place each chick into the brooder, offer a drink by dipping its beak into a bit of room-temperature sugar water, water with some Vitamins & Electrolytes mixed in, or plain Pedialyte. The chicks should have access to clean, fresh water around the clock in a shallow dish or chick-sized waterer with marbles or stones in it so they can't fall in and drown, or sit in it and get chilled.

Sprinkle some chick feed on a paper plate for them to peck at, and also fill a small

chick-sized feeder. Don't be surprised if the chicks don't eat right away, especially those you hatch yourself or buy day-old. Chicks generally don't eat anything for the first 48 hours or so, having absorbed the yolk in the egg they hatched from just prior to hatching.

Check on your chicks frequently until you are sure they are comfortable temperature-wise, and to be sure that they haven't knocked over their feeder or waterer. Clean and refill the feeder and waterer as needed. Chicks manage to kick shavings into both, so frequent cleanings will be necessary.

Chick feed is specially formulated to meet their dietary needs.

I provide feed and water to my chicks day and night until they are big enough to join the grown hens in the run, at which point they only get food and water during the day.

Water

Chicks should be provided unlimited access to clean, fresh water. A bit of sugar can be added to the water for the first few days to give your chicks a little energy boost. Water should be room temperature, not cold. I add a few drops of **apple cider vinegar** every few days to my chicks'

Chicks need clean, fresh water in specially-sized waterers. Small stones or marbles can help prevent accidental drowning.

water right from the start. The apple cider vinegar helps them build strong immune systems and aids with digestive health. It also kills the germs that can cause respiratory issues. Raw, organic apple cider vinegar with the mother is best (see my recipe for home-made apple cider vinegar on page 52).

Herbal tea

I give my baby chicks herbal tea from day one. Steeped herbs will help build strong immune systems and offer a wide array of health benefits for growing bodies without your worrying about the chicks not being able to digest tough plant fibers in the herb plants themselves. **Oregano** has been studied relative to use with commercial poultry and is thought to combat coccidia, E. coli and Salmonella, among other diseases that can afflict your chicks, so oregano tea is my favorite choice for baby chicks. I brew the oregano tea using either fresh or dried oregano, let it steep in hot water for ten minutes or so, then cool it to room temperature. For chicks, I don't make the tea their sole water source, but instead serve it alongside their plain water; this is a way for them to build resistance to various pathogens in the amounts they want or feel they need. Many other herbs provide varied health benefits for baby chicks, and all can be brewed into nutritious chick tea. See below for a more complete list of which herbs specifically benefit growing chicks.

Feed

In lieu of medicated chick feed, I choose to feed only **unmedicated feed**, available to them 24/7 for the first few weeks, and start my chicks on a regimen of fresh and dried herbs to not only boost their overall health and opportunity to fight pathogens themselves, but also to get them accustomed to the taste of the various herbs. Oregano, as mentioned above, is thought to combat and provide resistance to coccidiosis, which is the number one cause of death in chicks. Dried or fresh oregano in their diet is extremely beneficial in keeping your chicks healthy. **Probiotics** are an added guard against coccidia and other bad bacteria building up in your chicks' intestinal systems. Probiotic powder is a better choice than plain yogurt, which, while beneficial, can cause diarrhea if fed in excess. **Cloves** and **grapefruit seed oil** are also thought to help combat coccidia.

Grit. I provide the chicks a dish of coarse dirt and small stones (commercial "chick grit" is also available), which they need to help them grind and digest their food. If their diet includes anything other than chick feed, they do need the grit.

Chick treats. Just before a chick hatches, it ingests the yolk in the egg from which it hatches. Eggs contain every nutrient needed for life, except Vitamin C, and are therefore one of the most nutritious "treats" you can offer to your chicks.

A favorite of mine to feed my chicks is "egg custard." I whisk **eggs**, minced **garlic**, **dandelion greens**, a touch of **honey** and a bit of water over low heat in a frying pan until cooked, then let it cool and serve it to them in small amounts as a nutritious treat. Other good supplements to their diet are cooked or raw **oats**, **chickweed**, boiled **brown rice** and **chopped grass** and **weeds**. Treats should be limited and nutritious. A good-quality chick feed does provide the growing chicks all the nutrition they need, but I find that later in life they will be more apt to eat a wide variety of foods if they are introduced to them early as baby chicks.

Grass and weeds for good health

Chicks will enjoy plucking off short pieces of tender **grass** if you put a whole clump (grass, dirt and all) in their brooder. Just be sure any grasses or **weeds** you offer them are untreated with any pesticides or herbicides – and offer them whole, with the dirt attached. If you offer a whole clump, the chicks will nibble short pieces, preventing the crop issues that can possibly occur if you offer cut grass. The dirt doubles as grit for the chicks to help them digest the grasses and weeds.

Even more importantly, being exposed in small doses to the soil and grasses (in what is to become their eventual living environment) helps chicks build healthy immune systems and a natural resistance to the pathogens and bacteria in that

Tender grass and weeds are healthy for chicks.

environment. By putting the grass clumps in the brooder with them, you are bringing a small part of the outdoors inside to them, mimicking what they would face had they been hatched under a hen and followed her around the yard exploring the world. Chicks hatched under a mother hen would be out foraging for weeds and grasses at just a few days old, so the more you can replicate a "free range" environment in your brooder, the better for your growing chicks.

Grass and weeds also contain protein and other necessary vitamins and nutrients and are an easy, free, natural way to supplement your chicks' diet. Pecking in the dirt and looking for bugs and worms helps keep them busy and not interested in pecking at each other.

Fresh herbs for good health

I offer my chicks chopped fresh herbs every few days. Even if they don't eat them all, merely brushing against the oils can impart benefits. Several herbs in particular can have exceptional health benefits for growing chicks. **Parsley** helps with blood vessel development and circulation. **Violets** are good for circulation, too. **Bay leaves** help build strong immune systems, while **cilantro** and **comfrey** contribute to strong bones. **Pineapple sage** aids in nervous system development, and **nettles, parsley, sage** and **spearmint** are good for overall health. **Rosemary, dill, parsley** and **mint** assist in feather growth Respiratory health is very important to growing chicks. Both the ammonia in their excrement and the dust in their bedding can interfere with proper lung function, so adding some **basil, bee balm, cinnamon, clover, dill, echinacea, rosemary, thyme** or **yarrow** to the brooder – all of which aid in respiratory health – is extremely beneficial.

Protein plays an important role in your chicks' development, especially when they are growing in their feathers. Herbs high in protein include **basil, chervil, coriander, dill, fennel, marjoram, parsley, spearmint** and **tarragon**. Chopping these fresh and offering them in your brooder will result in stronger, healthier chicks.

Preventing B2 Deficiency

*Chicks that are deficient in Vitamin B2 can develop curled toes. To help prevent this deformity, which affects the nervous system, specifically the sciatic nerve – and which if not corrected can lead to paralysis, anemia and sometimes even death – be sure that your chicks have adequate amounts of riboflavin in their diet. Good sources include **brewer's yeast, wheat bran, sesame seeds, seaweed, spearmint, parsley** and **coriander**.*

Starting off your chicks as naturally as possible puts them on the path to a long, productive life. Adding apple cider vinegar, garlic, probiotics and various herbs to their diet early in their life is an easy way to turn healthy chicks into healthy hens.

HERBS IN THE BROODER

Bay leaves	Dill	Pineapple sage
Basil	Echinacea	Rosemary
Chervil	Fennel	Sage
Cilantro	Grapefruit seed extract	Spearmint
Cinnamon	Lavender	Tarragon
Clover	Marjoram	Thyme
Cloves	Nettles	Violets
Comfrey	Oregano	Yarrow
Coriander	Parsley	

THIRTEEN

WHEN SOMETHING IS WRONG

Using the recommendations presented throughout this book and taking pains to provide your flock a clean, safe place in which to spend their days and a secure coop at night, you should be able to mitigate most serious health problems and injuries. With diligent flock management – which includes no overcrowding, a dry, well-ventilated coop, a well-maintained run area, access to good-quality feed and clean, fresh water, and only bringing in healthy hens from reputable sources while instituting a strict quarantine period – sickness should be kept to a minimum within your flock. Add to that some herbs and supplements for strong immune systems and most health issues can usually be averted. There are bacteria and pathogens everywhere, especially in the area in which chickens live, but a strong, healthy chicken with a working immune system should be able to ward off most illnesses and parasites without a problem.

When we first started raising chickens, I made the conscious decision to use lots of natural preventives to help our hens build strong immune systems so they would be able to fend off any pathogens or germs they encountered without having to rely on commercial products. Many times, it's hard to find a vet who will treat chickens, and even if you find one who will agree to take a look at a sick hen, there are few medications that are approved for use on chickens. Working at raising healthy chickens is your best defense.

A strong immune system and lots of natural preventives go a long way toward keeping chickens healthy and able to fight off all sorts of parasites and pathogens successfully *on their own, from the inside.* This is so beneficial because it alleviates your need to try to identify, diagnose or treat issues. Instead, you're giving your hens the tools to heal themselves.

So far, my holistic approach seems to be working just fine. We have not had any trouble with parasites, sickness or unexplained deaths within our flock. Our chickens are gorgeous, vigorous and healthy – and lay beautiful eggs.

But even with the best intentions and proper practices, sometimes things go wrong. Even the best cared for chickens can get sick, and accidents happen. Some things are unavoidable and even genetic. My first instinct always is to reach for a holistic, non-chemical cure, but of course a vet should be consulted for a serious life-threatening injury, if what you are administering doesn't seem to be working after a reasonable period of time or if your hen's health continues to deteriorate and isn't responding to natural at-home treatment.

Deciphering symptoms: first steps

Chickens are masters at hiding signs of something being wrong; that is due to the pecking order instinct in which the weaker members of the flock can be unmercifully pecked, sometimes even to the death, or shunned from the flock. No chicken wants to appear weakened because that also makes her an easy target for a predator. But if you spend enough time around your flock you will learn to recognize normal behavior and be able to sense pretty quickly when something is wrong. Chickens have a very fast metabolism and their normal body temperature hovers between 102 and 107 degrees, so bacteria multiply much faster in their bodies than they do in the human body; and when illness hits, it progresses very quickly. Many times, by the time a chicken shows any symptoms at all, it's too late, but being in tune to subtle changes in behavior and appearance can help you catch and treat problems quickly.

General treatment

Listless hens should be immediately segregated, kept warm and started on a stepped-up regimen of natural remedies: a healthy dose of **apple cider vinegar** and **electrolytes** in their water, **probiotics**, and fresh **minced garlic**. Often, a few days in a warm, quiet place with the added nutrients will do the trick and your chicken will be back to her usual perky self in no time. If you have a hen who is hunched over, completely inactive, weak, coughing, sneezing or just looks terribly unhappy, it could be one of several serious infectious diseases and you should seek immediate treatment by a qualified vet. Go with your gut. You will know when something is seriously wrong.

First aid kit

Preparation is always important, so assembling a first aid kit in advance is a good idea since too often in backyard chicken keeping, unexpected injury or illness occurs that needs to be treated quickly. Hopefully, you will never need most of these items, but at least you'll be ready if you do.

I consider the items listed below essential to any chicken first aid kit. They are all-natural and have no side effects or withdrawal periods (the time you need to wait until you can safely eat their eggs again). Because we eat their eggs, and because many medications and treatments are not even approved for use in poultry – and withdrawal periods are often uncertain – I only use natural products.

A well-stocked first aid kit is essential.

Here's what I recommend for your chicken first aid kit:

- **Bach Rescue Remedy for Pets** – a natural stress reliever
- **Blackstrap molasses** – to induce diarrhea and flush toxins in case of accidental poisoning
- **Blu-Kote** – an antibacterial/antifungal spray for wounds
- **Calendula cream and/or coconut oil** – to prevent/treat frostbite on combs or feet
- **Cornstarch** – to stop bleeding
- **Epsom salts** – for soaking a foot with a splinter or bumblefoot infection (Also when ingested, Epsom salts can neutralize and help flush toxins, help with intestinal tract blockage, and reduce diarrhea.)
- **Honey** – natural antiseptic with healing properties
- **Kocci Free** – all-natural anti-parasitic and coccidiosis remedy
- **Liquid calcium** – helps an egg bound hen or hen laying soft-shelled eggs
- **Nutri-Drench** – liquid vitamins and other nutrients to boost the energy of a weak or ailing hen
- **Saline solution** – to rinse dirt or dust out of eyes or clean a wound
- **Verm-X** – an all-natural wormer
- **Vetericyn** – non-toxic spray that kills 99.99% of bacteria in cuts, scratches and bumblefoot infections
- **VetRx for Poultry** – 100% natural camphor-based formula used to treat respiratory ailments and eye worm
- **Vitamins & Electrolytes or plain, unflavored Pedialyte** – replenishes electrolytes during extremely hot weather or supplements an ailing hen
- **Vitamin E** – for the treatment of wry neck, a vitamin deficiency found mainly in chicks
- **Gauze pads, first aid tape, vet wrap, sharp scissors, Popsicle sticks, eyedropper**

In addition, you should keep a small pet carrier and a soft blanket nearby for a possible trip to the vet. It is also a good idea to keep a dog crate or large birdcage handy. It will make a wonderful "recovery room" for a chicken who needs to be separated for a while to recover from an injury or illness. That way, any special diet can be administered and eating/drinking/laying/pooping can be monitored, while keeping the chicken fairly immobile and safe from pecking from the others.

Keeping Yourself Healthy

While it is possible for humans to catch certain things from chickens, such as ringworm, mites or respiratory ailments from spores or other pathogens, as well as E. coli and Salmonella, using some common sense and keeping your own immune system strong will go a long way: Wear a dust mask when cleaning your coop or setting up a dust bath area; wash your hands after handling your chickens or working in their area; change your footwear before coming back into the house and be sure to change soiled clothing. Don't put your hands in your mouth, eat or drink in the coop area, and don't rub your eyes.

TREATING THE MOST COMMON PROBLEMS YOU WILL ENCOUNTER

Not much study has been done regarding herbal remedies and their use on chickens, but since they are inexpensive, easy to administer and, from my first-hand experience, extremely effective, they are my go-to choice for chicken health care. For example, **cinnamon** and **turmeric** are often used to fight bacterial growth. **Thyme** and **oregano** are being studied as natural antibiotic alternatives. **Chickweed** and **chamomile** are natural pain relievers. Depending on the symptoms you see, you might want to try some targeted herbal remedies first.

Respiratory issues

The vast majority of chicken illnesses are respiratory in nature. Chickens are very susceptible to issues affecting their breathing. They have elaborate respiratory systems and small, weak lungs, which can lead to problems that are difficult to diagnose and treat. Wheezing, watery eyes and nose, squeaking noises, sneezing or coughing are signs of respiratory problems.

However, often what appears to be respiratory distress is nothing more serious than dust or other debris in the eyes or sinus cavities, or something lodged in the throat. Even rough mating can affect the sinuses. Try massaging the throat and giving the hen a drink of **olive oil**. A squirt of **saline solution** in each eye several times a day can help ease the symptoms simply by clearing the sinuses. Administering fresh **garlic, apple cider vinegar** and **yarrow** is also beneficial.

Cinnamon, basil, thyme and dill are just a few of the herbs used for treating the respiratory system.

Basil, cinnamon, clover, dill, echinacea and **thyme** all have positive effects on the respiratory system. Try chopping up the fresh herbs (or using dried, if fresh is not available) and sprinkling a bit of ground **cinnamon** on top, then offering it to your ailing hen, free-choice or mixed into her feed.

Respiratory issues can also signal infectious bronchitis or coryza. Symptoms that don't clear up in a week or so, or continue to get worse despite your treatment, could signal a potentially serious illness, and the advice of a veterinarian should be sought.

Eye problems

Cloudy eyes, watery eyes and beak or rubbing of the eyes can signal conjunctivitis, which can result from a buildup of ammonia in the coop. Again, flush the eyes with **saline** and change out all the bedding in your coop. Try drops of cooled **chamomile tea** in each eye (or press a steeped and cooled chamomile tea bag to the eye area) several times a day until the eyes clear up.

Goldenseal soothes inflamed mucous membranes. Two teaspoons dried goldenseal steeped in 8 ounces of hot water for 15 minutes and then applied to the sinuses and dripped into the nostrils or eyes can also help with conjunctivitis or sinusitis.

Swollen, pus-filled eyes, or eyelids that are stuck together can be signs of *eye worm*. Sometimes the worms are even visible under the lids, swimming around. (Sparing you all the gross details, it's basically a worm contracted from cockroaches.) Your chicken will begin to scratch at her eye with the tip of her wing and could literally scratch her eyeball out. **VetRx** is my recommended treatment for eye worm.

Sour crop

Sour crop is the potentially fatal condition that occurs when a hen's crop does not fully empty and a bacterial yeast infection results. The following can cause sour crop: inadequate access to grit; consuming long, fibrous grass, weeds and other plant fibers; excessive amounts of bread products or pasta; eating wood chips, sand, string or twine; moldy feed.

Sour crop is identified by a squishy, soft, engorged crop, sour breath and possibly some liquid coming out of the hen's mouth. To treat sour crop, massage the crop in the direction of the head to try to induce vomiting, and feed the hen **yogurt, olive or vegetable oil and water with apple cider vinegar** in it, along with additional **grit** to aid in digesting the mass. Withhold all other solid foods and continue to massage the crop gently several times a day until the condition improves.

Preventive measures include providing plenty of grit and adding apple cider vinegar to their water several times a week. Free range chickens rarely suffer sour crop since they nip short pieces of grass; so if you hand pick grass and weeds for your chickens, be sure to chop them first.

Impacted crop

Impacted crop is a related but slightly different problem, which is indicated by a hard crop and possibly trouble breathing. An impacted crop presses against the windpipe and can literally suffocate the hen. Caused by large pieces of food in the crop that can't pass through the digestive system – or if a chicken inadvertently swallows pieces of metal, plastic, rubber bands or other indigestible substances – impacted crop can be treated by orally administering some **olive or vegetable oil** mixed with warm water and then massaging the crop every few hours to try and break up the blockage. Offering **grit** to try and break up the impaction can also help; and again, solid foods should be withheld until the crop has emptied. In extreme cases, it might be necessary to slit the crop open with a scalpel and remove the blockage.

Egg bound hen

Your chicken might be egg bound (meaning that an egg is stuck inside her) if she is fluffed up and her eyes are closed. You may notice her sitting on the ground and maybe dragging her wings, tail down. She may also be straining or pumping her backside.

Egg binding could be due to a large or double-yolked egg that is too large to pass through the oviduct; or it could be due to genetics or a calcium deficiency. Calcium is needed for proper muscle contraction. Too much protein in a hen's diet can also cause egg binding. Other potential causes are internal worms, low quality feed, dehydration or weakness from a recent illness.

You want to handle your egg bound chicken carefully to avoid breaking the egg inside her, which can lead to infection, peritonitis or possibly death. Even if the egg is not broken, the condition must be treated quickly. An egg bound hen will die if she is not able to pass the egg within 48 hours, so once you have made your diagnosis, treatment should start immediately.

To treat an egg bound hen, soak her in a plastic tub in warm water with some **Epsom salts** for about 20 minutes, then gently towel dry or blow dry her on low heat (yes, it is not only possible to blow dry a chicken, they actually seem to enjoy it).

Then rub some **vegetable oil** around her vent and very gently massage her lower abdomen. Put her in a quiet, dark location in a large crate or cage. Drape a towel or blanket over it; on the bottom, lay a towel that has been warmed in the dryer. Give her 1cc of **liquid calcium** with an eyedropper. Sprigs of **fresh lavender** in the cage or a spritz of **lavender essential oil** misted into the air might help her relax even more. Then give her some time to herself.

Repeat the soak and massage every hour or so until she lays her egg. After several tries, if she hasn't laid an egg and seems unable to relax, you might try putting her into a nesting box in the coop, where some

A warm Epsom salt bath is the first step to treating an egg bound hen.

fresh lavender sprigs might help as well. Calming down is often what she needs to finally lay her egg. Use your judgment on this one, but if being indoors in a crate seems to be really stressful to her, returning her to familiar surroundings might be the answer.

As a last resort, a visit to a vet is recommended; or, if you can see the egg, you can carefully extract the contents of the egg with a syringe and then gently crush the shell, keeping the fragments attached to the membrane. Remove the membrane and fragments as you squirt vegetable oil in and around the vent.

Vent prolapse

Occasionally, a hen's vent prolapses. This condition occurs when a portion of her oviduct ends up outside her body. It is serious and likely to happen on a recurring basis after it has occurred once. It is more common in chickens forced to lay through the winter instead of being able to adhere to a natural laying cycle, overweight hens, young hens or hens who lay extremely large eggs or soft-shelled eggs.

Treatment should include withholding feed for at least 24 hours followed by several days of light rations of mainly greens and some milk, plus clean water to prevent another egg being laid. Carefully cleaning the prolapse and then pushing it back into place before applying **witch hazel** is the recommended treatment. Witch hazel is a natural anti-inflammatory which will help tighten the skin and vent, as well as reduce the pain and swelling. The treatment should be repeated daily until the prolapse has healed; the hen should be kept separate for the duration to prevent pecking by flock members. Vigilance is mandatory to be sure the vent stays in place.

Coccidia/Coccidiosis

Coccidiosis, a microscopic parasite, is the number-one cause of death in baby chicks. Lethargy and bloody stools are two common signs your chicks may be infected. Although older hens can also contract coccidiosis, they generally are strong enough and have built up a tolerance to the low levels normally found in the run. Most chicken runs, no matter how clean or well kept, contain traces, and the best defense against it is small, regular exposure so the body can build up its own immunity.

As I mentioned, some commercial chicken farms are beginning to test **oregano oil** and **cinnamon** as natural antibiotics and immune system boosters. Here are some other natural remedies that can help: fresh minced **garlic** to give their immune systems a nice boost and **green tea** and **safflower oil** to help control coccidia and strengthen the immune system. A bit of **apple cider vinegar** splashed in their water and some **probiotic powder** in their feed can help combat intestinal problems in chicks, as well as supporting immunity.

Oregano is one of my go-to herbs for treating chickens.

Lactic acid can also help flush out the parasite. Try a mash of equal parts feed and milk, mixed with some plain yogurt. Since chickens are unable to digest the milk sugars in dairy products, this causes diarrhea. Not normally a desirable thing, but in the case of a parasitic illness, the diarrhea works to cleanse the intestines.

Salmonella

Salmonella (or Salmonellosis) is a bacteria affecting the intestinal tract of humans, chickens and other birds and mammals. Symptoms in chickens include weak and lethargic birds, loose yellow or green droppings, purplish combs and wattles, a drop in egg production, increased thirst, decreased feed consumption and weight loss. It can be deadly in hens if not treated. Salmonella is usually spread to chickens through rat or mouse droppings in water, feed, damp soil or bedding/litter.

Sage is valuable for treating Salmonella.

Sage and **oregano** have been found to be effective in combatting Salmonella. Fresh or dried, these herbs can be added to your flock's diet on a regular basis as a preventive as well.

E. coli

E. coli is the collective name for a group of diseases that cause diarrhea and other infections in chickens. Mainly spread by infected wild birds and rodents in their droppings, E. coli can be largely prevented with proper biosecurity practices and clean runs, feeders and waterers. **Oregano** and **probiotics** are thought to mitigate the effects of an E. coli infection.

Diarrhea

There is a wide range of "normal" when it comes to chicken poop. Although diarrhea can signal illness, often in the summer when water intake is increased, droppings will normally become runny.

Some natural ways to treat diarrhea include mixing some **plain yogurt, brewer's yeast, charcoal powder** and **dill** and feeding it to the afflicted hen. **Boiled rice, honey, milk** and grated **raw apple** can also help, as can adding **probiotic powder** to the feed to alleviate future bouts.

Provide fresh, clean water since diarrhea dehydrates the chicken and she will need to replenish the liquid lost. Adding **Vitamins & Electrolytes** to the water is also beneficial for her recovery.

Botulism & accidental poisoning

Botulism poisoning can be contracted by hens eating spoiled or rotting food or drinking water from contaminated ponds or puddles. Symptoms of botulism poisoning include respiratory distress, diarrhea, twitching, loss of muscle control and eventually paralysis.

In case of accidental poisoning (from antifreeze, fertilizer, mothballs, insecticides, etc.) or suspected botulism poisoning, a **molasses flush** can help by absorbing the toxins and flushing them out of the body. The recommended treatment is adding three tablespoons of blackstrap molasses per gallon of water and offering it as the sole source of drinking water for eight hours – then offering fresh, clean water. You can follow that treatment up with a molasses cleanser to reintroduce good bacteria: Puree several cored apples, 1 egg yolk, 1 teaspoon blackstrap molasses and ¼ cup plain yogurt in a food processor. Feed free-choice.

Internal parasites: including gapeworm, tapeworm, threadworm and roundworm

Various types of internal worms can be transmitted to your chicken if she eats an infected earthworm, slug or insect. For the most part, healthy chickens can fend off serious infestations – their immune system will create an inhospitable environment for the parasites and naturally flush them – but weak or otherwise unhealthy hens can succumb to the worms. **Diatomaceous earth (DE)** can help combat parasites, both internal and external, as will dried, crushed **artemisia (wormwood), feverfew, oxeye** or **Shasta daisy** added to your flock's diet.

Internal worms and parasites are most active during a waxing full moon, so treating for them several days before a full moon is recommended for maximum effectiveness.

NATURAL WORMERS/WORM PREVENTIVES

Pumpkin seeds (as well as the seeds of other members of the cucurbitaceae family, such as **winter, summer, zucchini** and **crookneck squashes, gourds, cucumbers, cantaloupe** and **watermelon**) are coated with a substance called cucurbitacin, which paralyzes worms. The larger fruits and vegetables contain higher levels of cucurbitacin, while the smaller ones, such as cucumber, contain far less. Preventive wormings should be done twice a year, spring and fall. I make a slightly different recipe for each time of year using seasonal ingredients.

Pumpkin Seed Worming Paste for the Spring Worming

Serves 10-12 hens

What you'll need:

1 bulb fresh garlic
1 cup raw, unsalted, unshelled
   pumpkin seeds
1 cup raw old-fashioned oats
1 cup dandelion greens
1 cup nasturtium flowers
2 tablespoons blackstrap molasses
1 cup water

What to do:

Blend all ingredients in your food processor until the seeds are finely chopped and the mixture has reached a "paste" consistency. Feed free-choice or mix into scrambled eggs or oatmeal.

Makes approximately 2 cups.

Pumpkin Soup for the Fall Worming

Serves 10–12 hens

What you'll need:

1 bulb fresh garlic
2 cups raw old-fashioned oats
1 shredded carrot
Seeds and pulp from 2 medium-sized pumpkins (save the halved pumpkin shells)
2 tablespoons blackstrap molasses
Plain yogurt

What to do:

Blend all ingredients in the food processor and add enough plain, unflavored yogurt to achieve a "soupy" consistency. Pour some soup into each pumpkin half (or a bowl) and serve.

Here's a wonderful year-round natural wormer: Give your hens **garlic**, especially in conjunction with **mint, wormwood** and **nasturtium leaves and flowers**. In addition to being a natural wormer, nasturtium is also a laying stimulant, antiseptic and antibiotic. Year-round access to these natural wormers, along with the twice-yearly treatments, should keep your flock parasite-free.

Nasturtium is a natural wormer.

External parasites

Chickens can be susceptible to various external parasites as well, including lice, ticks, fleas and several types of mites. Access to a dust bathing area is the best way to prevent these parasites from becoming a problem, but if you find your chickens laden with tiny, crawling bugs, here are some natural remedies to try:

Lice. Tiny, straw-colored bugs that appear around the vent and under the wings, lice will lay eggs at the base of the feathers; the eggs look like white residue. Dusting the affected hen's vent area with food-grade **diatomaceous earth** can take care of the lice quickly and easily if you catch them early on. However, if the infestation has taken hold, here's one way to rid your flock of the lice:

> *Set up two plastic dish tubs. Mix 2 cups of salt into 2 gallons of warm water in one tub. Soak the hen for 10 minutes, then soak her for 5 minutes in the second tub full of 2 gallons of warm water with ½ cup Dawn dishwashing liquid and ½ cup white vinegar mixed in. The lice should drown and float to the top. Rinse her well and dry her thoroughly. Dust with food-grade diatomaceous earth. Repeat every other day or so until you no longer see evidence of the lice.*

Mites and lice are just a few parasites that chickens are susceptible to.

Mites. Mites are tiny, crawling external parasites that can become a problem for your chickens if they are given the opportunity to move in to your coop and take up residence on your flock. They are generally spread by bringing infected birds into your flock, by wild birds, rodents, or by carrying them in on your shoes or clothing. They tend to congregate around the chickens' vent area and look like black specks. You also might see evidence of mites on the roosts, as blood spots, if you wipe your hand along the roosts. Treating mites can be difficult, and the hens as well as your coop should be treated simultaneously, and on an ongoing basis over the course of several weeks, to completely eradicate them.

It's good practice to make your coop an inhospitable environment for the mites to infiltrate. They aren't fond of strongly aromatic scents, so…

Mite Preventives:

- Feed hens raw minced **garlic** free-choice or in their water. It's thought that blood-sucking parasites don't like the taste of blood when the host has garlic added to their diet.

- Mix **garlic powder, dried oregano, parsley, wood ash, rosemary** and **lavender** and rub into the skin.

- Tie bouquets of **wormwood** (artemisia), **basil, catnip, cayenne pepper, garlic, oregano, rosemary** and **sage** to the roosts or make sachets for your nesting boxes as an ongoing mite repellent.

Mite Treatment: A garlic juice mite spray mixture has been found by poultry scientists in the U.K. to have a 100% kill rate over 24 hours. This can be used as a treatment and also as an ongoing preventive (see recipe on next page).

Fresh or dried herbs such as lavender, oregano and rosemary are natural treatments for mites.

Natural Garlic Juice Mite Spray

What you'll need:

10 ounces water

1 ounce garlic juice

1 teaspoon (total) any combination of these essential oils:
   bay, cinnamon, clove, coriander, lavender, spearmint, tea tree and/or thyme

What to do:

Mix in a spray bottle and spray hens – bi-weekly as a preventive or every other day for two to three weeks in the case of an infestation. Concentrate around the vent and under the wings. To completely get rid of the mites, you will need to treat your coop and chickens simultaneously since some types of mites do leave the hens and burrow into wood or crevices.

Spraying your coop and roosts for several days in a row with a mixture of 2 cups water, 1 cup cooking oil and 1 tablespoon dishwashing liquid will help kill off the remaining mites by suffocating them. The coop should continue to be sprayed at least once or twice a week for several weeks.

While your chickens are suffering from mites, it is recommended you increase their iron intake to prevent anemia. Good sources of iron include: **eggs, meat, poultry, fish, seafood, spinach, beet greens, dandelion greens, sweet potato, broccoli, collards, kale, strawberries, watermelon, raisins, wheat products, oatmeal, cornmeal,** and **molasses**. These foods, added to their diet, can help them better battle the mites. In addition to draining the body of iron, mites also affect the immune system..

Scaly Leg Mites. The scales on the legs of healthy chickens are smooth and lie flat. If the scales on your chickens' feet start to peel up, flake or look rough and uneven, they could be suffering from leg mites. Soak their feet and legs in warm water, then scrub with an old toothbrush dipped in a **white vinegar/garlic juice mixture**. Let dry then apply some olive oil mixed with some **cayenne pepper** and a few drops of **tea tree oil** or **orange essential oil** every day, until the crust and old scales fall off.

Bumblefoot

Bumblefoot is a staph infection that starts with a cut foot or a hard landing from a roost and can travel up the leg. It is easily identified by the telltale black scab on the underside of a hen's foot. Treatment includes **calendula cream** or **Vetericyn** applied until the scab disappears. Cutting the foot with a scalpel and removing the scab and hard kernel inside the foot pad is a last resort if the natural topical treatment doesn't work.

Doing everything you can to keep your chickens' immune systems in top working order is paramount to flock health, and when treatment is needed, try a natural "cure" first.

Preventing Bumblefoot

*Bumblefoot is thought be partially caused by a biotin (Vitamin H) deficiency. Increasing your flock's biotin intake in their diet can be a natural bumblefoot preventive. Some foods that chickens enjoy which are good sources of biotin include: **Swiss chard, kale, cabbage, cucumbers, broccoli, cauliflower, sweet potatoes, nuts, cooked eggs, whole grains, oats, tomatoes and raspberries.***

Adding brewer's yeast to their daily feed is also an excellent way to ensure your flock is getting enough biotin.

WHEN SOMETHING IS WRONG

Apple cider vinegar	DE	Probiotic powder
Basil	Echinacea	Pumpkin seeds
Blackstrap molasses	Feverfew	Oregano
Catnip	Garlic	Sage
Cayenne pepper	Goldenseal	Tea tree oil
Cinnamon	Honey	Thyme
Daisy (oxeye or Shasta)	Lavender	Yarrow

Appendix

SAFE-TO-EAT WEEDS AND FLOWERS

Chickens love to eat weeds and flowers. When I let our flock out to free range in the pasture, they eat a fair amount of tender grass and are always on the lookout for bugs and worms – but they overwhelmingly seem interested in eating weeds and flowers. Most weeds are extremely nutritious and contain tons of vitamins, minerals and other nutrients. Winter weeds are especially good, healthy treats when grass and other greens are scarce.

Certain weeds and flowers – like marigolds, pennycress and alfalfa – contain the compounds that make your egg yolks bright orange; and some – like chickweed, henbit and fat hen – were originally so named because chickens loved them so much. Chickweed is also a natural pain reliever. Clover is one of the most nutrient-complete weeds you could feed your chickens, although it is a blood thinner so it should not be fed exclusively or in unlimited amounts.

Here are some of the more common edible weeds and flowers that grow in many parts of North America. They are all perfectly safe to feed to your chickens in unlimited amounts.

Apple blossoms	Echinacea	Mouse ear chickweed	Roses
Beautyberry (callicarpa)	Evening primrose	Mugwort	Shiny cudweed
Bee balm	Fat hen	Nasturtium	Smartweed (heart's ease/lady's thumb)
Bittercress (shotweed)	Geranium	Nettle	Snap dragon
Burweed	Hawkweed	Oxalis (yellow wood sorrel)	Squash blossoms
Calendula	Henbit	Pansy	Sunflower
Carnations	Hibiscus	Pea blossoms	Violets
Catchweed bedstraw	Hollyhock	Pennycress	Wild Carolina geranium
Catsear	Impatiens	Peony	Wild carrot (Queen Anne's lace)
Chickweed	Lavender	Phlox	Wild strawberry
Citrus blossoms	Lilac	Purple deadnettle	Yarrow
Dandelion	Marigold (calendula)	Purslane	

Plants to Steer Clear of:

Most of the common "yard" weeds are fine; for the most part, chickens will avoid eating things that are harmful to them, but to be on the safe side it's best to avoid deliberately planting the following potentially toxic plants and flowers in areas your chickens can access:

Azalea	Henbane	Philodendron
Black nightshade	Hydrangea	Privet
Buttercup (Ranunculus spp.)	Honeysuckle	Rhododendron
Castor bean	Irises	St. John's wort
Clematis	Lantana	Sweet pea
Corn cockle	Lily of the valley	Trumpet vine
Foxglove	Oleander	Vetch

There are many, many more potentially harmful plants...but if you stick to the safe AND nutritious choices listed on the previous page, your chickens will thank you!

HEALTH BENEFITS OF COMMON HERBS, WEEDS & FLOWERS

Basil – antibacterial, mucous membrane health, insecticide, rejuvenating

Bay leaves – antiseptic, antioxidant, immune system booster, insect repellent

Bee balm (Bergamot/Monarda) – antiseptic, antibacterial, respiratory health, calming, anti-diarrheal, digestive tract health

Burdock root – antioxidant, balancing

Calendula – anti-inflammatory, antifungal, controls bleeding, aids in healing wounds and skin ailments, pain reliever

Catnip – sedative, insecticide, lice repellent, rodent repellent, wormer

Cayenne pepper – aids circulation, appetite stimulant, antiseptic, digestive enhancement

Chamomile – calming, relaxant, wormer, provides calcium, aids digestion

Chickweed – pain reliever

Cilantro – antioxidant, anti-fungal, builds strong bones, high in Vitamin A for vision and Vitamin K for blood clotting

Cinnamon – promotes healthy breathing

Clover – antioxidant, boosts respiratory health and blood flow, detoxifier, aids in digestion

Comfrey – aids digestion, anti-inflammatory, supports bone and artery growth, heals sprains and broken bones, only known plant source of Vitamin B12, protein-rich

Dandelion greens – antioxidant, excellent source of calcium

Dill – antioxidant, relaxant, respiratory health, anti-diarrheal

Echinacea – aids in respiratory health, strengthens immune system

Fennel – laying stimulant

Feverfew – fly and flea repellent, anti-inflammatory,

Garlic – laying stimulant, anti-fungal, benefit circulatory system

Ginger – stress reducer, appetite stimulant, anti-oxidant

Lavender – stress reliever, increases blood circulation, highly aromatic, insecticide

Lemon balm – stress reliever, antibacterial, highly aromatic, rodent repellent, calming, insecticide, rejuvenating

Lemongrass (Citronella) – fly, mosquito and other flying insect repellent

Marigold – vibrant egg yolks, feet and beaks/bills, insect repellent, antioxidant, soothes irritated skin

Marjoram – laying stimulant

Mint (all kinds) – insecticide and rodent repellent

Nasturtium – laying stimulant, antiseptic, antibiotic, insecticide, wormer

Nettles – increases egg production, overall health boost, blood cleanser, wormer, high in vitamins and nutrients

Oregano – anti-parasitic, antifungal, antibiotic properties that are thought to combat coccidia, salmonella, infectious bronchitis, avian flu, blackhead and E. coli

Parsley – high in vitamins, especially the B vitamin choline, aids in blood vessel development, laying stimulant

Peppermint – anti-parasitic, insecticide

Pineapple sage – aids nervous system, highly aromatic

Rose petals – highly aromatic, high in Vitamin C, antiseptic, antibacterial, cleanse blood toxins

Rosemary – pain relief, respiratory health, insecticide, calming

Sage/Pineapple sage – antioxidant, antiparasitic, general health promoter, insecticide, calming

Slippery elm bark – antidiarrheal, soothes mucous membranes, wound healer

Spearmint – antiseptic, insecticide, stimulates nerves, brain and blood functions, eases fatigue

Squash blossoms – good source of calcium, iron and Vitamin A

Tarragon – antioxidant

Thyme – antibiotic, respiratory health, antibacterial, antioxidant, antiparasitic, hopping and crawling insect repellent

Violets – aid circulatory system, anti-inflammatory

Wormwood (Artemisia) – wormer, antitoxin, insecticide, antibacterial, antiseptic (use sparingly with caution)

Yarrow – antibacterial, anti-inflammatory, clears sinuses and respiratory systems, stress reliever, wound healer, insecticide

BY BENEFIT

Antibacterials – Basil, Bee balm, Garlic, Lemon balm, Roses, Thyme, Wormwood, Yarrow

Antibiotics – Nasturtium, Oregano, Thyme

Anti-diarrheal – Bee balm, Dill, Slippery elm bark

Antioxidants – Bay leaves, Bee balm, Burdock root, Cayenne, Cilantro, Clover, Dandelion greens, Dill, Ginger, Marigold, Roses, Sage, Tarragon

Antifungals – Calendula, Cilantro, Garlic, Oregano

Anti-inflammatories – Calendula, Comfrey, Feverfew, Violets, Yarrow

Antiparasitics – Oregano, Peppermint, Sage, Tarragon

Antiseptics – Nasturtium, Roses, Spearmint, Wormwood

Antitoxins – Clover, Rose Petals, Wormwood

Appetite stimulants – Cayenne, Ginger

Aromatics – Lavender, Lemon balm, Pineapple sage

Blood clotting – Calendula, Cilantro

Blood vessel development – Parsley

Bone strength – Cilantro, Comfrey

Calming – Bee balm, Catnip, Chamomile, Dill, Ginger, Lavender, Lemon balm, Rosemary, Sage, Yarrow

Circulation – Cayenne, Garlic, Lavender, Violets

Digestive health – Bee balm, Cayenne, Chamomile, Clover, Comfrey, Ginger

Disease combatants (coccidiosis, Salmonella, infectious bronchitis, avian flu, blackhead, E. coli) – Oregano

Egg yolk, foot and beak color – Marigolds

Fly repellents – Basil, Dill, Feverfew, Lemongrass, Mint, Rosemary

Healing – Calendula, Comfrey, Yarrow

Immune system boosters – Bay leaves, Echinacea

Insecticides – Basil, Bay leaves, Catnip, Lavender, Lemon balm, Lemongrass (Citronella), Marigold, Mint, Nasturtium, Rosemary, Sage, Thyme, Wormwood, Yarrow

Laying stimulants – Fennel, Garlic, Marjoram, Nasturtium, Nettles, Parsley

Lice repellent – Catnip

Nervous system aids – Pineapple sage

Overall health – Nettles, Parsley, Sage, Spearmint

Pain relievers – Calendula, Chickweed, Rosemary

Rejuvenating – Basil, Lemon balm

Respiratory health – Basil, Bee balm, Cinnamon, Clover, Dill, Echinacea, Rosemary, Slippery elm bark, Thyme, Yarrow

Rodent repellents – Catnip, Lemon balm, Mint

Skin tissue growth – Marigolds

Wormers – Catnip, Chamomile, Garlic, Nasturtium, Nettles, Wormwood

Wound healing – Calendula, Slippery elm bark, Yarrow

ACKNOWLEDGMENTS

I want to thank my husband for indulging my "whim" to start a small backyard chicken flock and then not protesting (too loudly at least) when that flock grew to more than 30 chickens, a rooster, 11 ducks – and, most recently, a corgi to herd them all. Thank you for always being here to give your honest opinion and to fix all my computer problems (and not yell at me when it really IS my fault that things aren't working right). You always encourage me to do the right thing and be true to myself, and I can't thank you enough for that.

Thanks to my mom for teaching me that I really can do anything I put my mind to – and do it well. You have been an inspiration to me my whole life and I love you for many reasons, but mostly because you will always support me no matter what I decide to tackle.

A very special thank-you to Yvette and Suzanne for not only providing moral support, but also making me laugh every day without fail: To Yvette for keeping me focused and organized, and to Suzanne for helping me keep a level head and reminding me that nice girls DO finish first. Thank you both for picking up the slack and keeping our Facebook page hopping to allow me time to write this book. I consider you both among my very best friends and I know that one day we will meet in person.

Thanks to each and every Fresh Eggs Daily fan who has shared in this journey and in many ways made this book possible. I hope you enjoy it!

And a huge thank-you to everyone at St. Lynn's Press – especially Paul, Cathy, Holly and Marguerite – for taking a chance and believing in me and my idea for a book about chickens and for making the entire process so enjoyable. Let's do it again sometime!

INDEX

Recipes and Projects

References and Further Reading

Backyard Poultry – Naturally, by Alanna Moore, ACRES USA, 2007

The Complete Herbal Handbook for Farm and Stable, by Juliette de Bairacli Levy, Faber & Faber, 1991

Folk Medicine, by D.C. Jarvis, M.D., Holt, Rhinehart and Winston, 1958

The Herb Book, by John Lust, Dover Publications, 2014 (the latest reprint of this classic 1974 work)

Kitchen Medicines, by Ben Charles Harris, Random House, 1968

ABOUT THE AUTHOR

LISA STEELE lives on a Virginia farm with her husband and an ever-growing number of chickens, ducks and assorted four-legged friends. Her happy, healthy animals benefit from her wide knowledge of herbs and other natural health enhancers and preventives. Lisa blogs about life as a farm girl and raising chickens and ducks on her popular website Fresh Eggs Daily (www.fresh-eggs-daily.com). She also shares on her three Facebook pages: Fresh Eggs Daily, Ducks Too, and A Farm Girl and her Chickens.

Since 2009, when she first "hatched" the concept of Fresh Eggs Daily at the kitchen table of her farm, Lisa has been featured in numerous magazines, including *Hobby Farm Home*, *Hobby Farm Chickens*, *Your Chickens*, *Country Woman*, *Good Housekeeping* and *Southern Living*. Her writing has appeared in HGTV Gardens and *Grit Magazine* online.

In her free time, Lisa enjoys hauling hay and fixing fences around the farm with her husband, spending time with her flock, horses, two dogs and cat, cooking for friends and family using fresh herbs and vegetables from her garden, and baking with eggs fresh from her coop. She is also an avid reader and knitter. Lisa invites you to contact her at Fresheggsdaily@gmail.com.